T0257623

Maintaining Effective Engineering Leadership

A new dependence on effective process

Other volumes in this series:

Maintaining Effective Engineering Leadership

A new dependence on effective process

Raymond Morrison

The Institution of Engineering and Technology

Published by The Institution of Engineering and Technology, London, United Kingdom

The Institution of Engineering and Technology is registered as a Charity in England & Wales (no. 211014) and Scotland (no. SC038698).

© The Institution of Engineering and Technology 2013

First published 2013

The Institution of Engineering and Technology
Michael Faraday House
Six Hills Way, Stevenage
Herts, SG1 2AY, United Kingdom

www.theiet.org

British Library Cataloguing in Publication Data
A catalogue record for this product is available from the British Library

ISBN 978-1-84919-689-5 (hardback)
ISBN 978-1-84919-690-1 (PDF)

Typeset in India by MPS Limited
Printed in the UK by CPI Group (UK) Ltd, Croydon

Contents

Foreword

I believe that the purpose of this book is to provide the aspiring engineering leader with an understanding of the criteria required to be an exceptional leader in his or her group or company. The difficulties in the current world of work are the high-level requirements and complexity of the work itself that manifests the functions in the projects, programs, and services that must be provided. The requirement is only a piece of what the engineer must endure in putting together the overall project. The four components of the ELITE Leadership Model provide the reader with most of the tools to complete their functions and requirements in a more orderly manner. In addition, the other tools that are presented in this text will help the aspiring engineering leader as well to see the disorganized world in a more organized way and to provide the means for attacking the problems with more wholesome ideas of resolve.

The author begins this book with a look at the "Columbia" disaster and what may have gone wrong. In themselves the NASA programs were very complex systems that required large crews of people to organize and coordinate the activities. With that promise comes the obvious confusion. It is believed that there is some resolution for the reader to be had by following the ideas presented in this volume.

Acknowledgements

Without the experiences and encouragement of the administrators (Pat Hall and Nancy Kruse) and students of the ELITE Leadership Program at the University of Tulsa, Oklahoma, USA, this book would not have been possible. The lessons learned and the experiences attempted (both successful and not so successful) were all contributors to the learning and understanding, which has developed in this text, making the publication that more possible.

In addition, my most sincere thanks goes to my wife, Christine Morrison, for her understanding and patience with my moods and devotion to getting this done. Her support in the development of this project and the encouragement are greatly appreciated.

I am also grateful for the encouragement of John Lorriman in the UK and other members of the ASEE in the United States, who provided input and encouragement at the various meetings of the College Industry Education Conferences and other national meetings to those who pushed me to write this book.

Paul Deards and his team of publishers at the IET were most helpful in their understanding and appreciation for the need of this publication, their help is greatly appreciated. It was through all of the discussions and interviews that were conducted that the pieces fell into place, thanks to each of you.

Chapter 1

A good process gone bad: setting the stage with the *Columbia* disaster

Figure 1.1   Shuttle Columbia blasting off from Kennedy Space Center
            [Source: NASA]

There are few people in the world today who are not familiar with the disaster that consumed the Space Shuttle *Columbia* on its return to earth on 1 February 2003. It captivated the world's television viewing population at the time and was probably one of those events that will be remembered in memory snapshots of where you were or what were you doing when the 'Columbia disaster' was announced.

NASA reported:

Columbia re-entered Earth's atmosphere with a pre-existing breach in the leading edge of its left wing in the vicinity of Reinforced Carbon-Carbon (RCC) panel 8. This breach, caused by the foam strike on ascent, was of sufficient size to allow superheated air (probably exceeding 5,000 degrees

Figure 1.2   Columbia to orbit [Source: NASA]

Fahrenheit) to penetrate the cavity behind the RCC panel. The breach widened, destroying the insulation protecting the wing's leading edge support structure, and the superheated air eventually melted the thin aluminum wing spar. Once in the interior, the superheated air began to destroy the left wing. This destructive process was carefully reconstructed from the recordings of hundreds of sensors inside the wing, and from the analyses of the reactions of the flight control systems to the changes in aerodynamic forces. ... By the time Columbia passed over the coast of California in the pre-dawn hours of February 1, at Entry Interface plus 555 seconds, amateur video show that pieces of the Orbiter were shedding. ... Analysis indicates that the Orbiter continued to fly its pre-planned flight profile, although, still unknown to anyone on the ground or about Columbia, her control systems were working furiously to maintain that flight profile. Finally, over Texas ... the denser levels of the atmosphere overcame the catastrophically damaged left wing, causing the Orbiter to fall out of control at speeds in excess of 10,000 mph. [1]

Seven astronauts were tragically lost. It must be clarified that this book does not intend to point fingers or find blame (Figure 1.3). As an author interested in the use of tested processes, I only want to examine the processes used in NASA at the time of the disaster; I will later generalise them as they relate to all types of organisations. These ideas consider the problems which occur when processes are set aside or changed because of some outside influence or assumption of immediate or assumed

Figure 1.3   Columbia crew [Source: NASA]

need. It is worth examining why these assumptions inform the management decisions made under the circumstances and what immediate consequences may result. It
is the current complexity of the changes to effective processes and the subsequent
difficulties which these actions produce for those involved or affected, this is the
focus. *Columbia* was selected as a jumping off point, not because of NASA's problems, but because of their visibility in the world of complex systems and the
resulting very complex management decisions that were made.

By their very nature, high-risk technologies are exceptionally difficult to
manage. Complex and intricate, they consist of numerous interrelated
parts. Standing alone, components may function adequately, and failure
modes may be anticipated. Yet when components are integrated into a
total system and work in concert, unanticipated interactions can occur that
can lead to catastrophic outcomes. [5]

1.1   Effective use of process: the name of the game, this occurs in more than just the space race and the Shuttle Program

Before any craft leaves its berth, its captain makes a detailed checklist walk around and about the equipment, to assure that everything is 'shipshape' and that everyone has done their job to assure a sound vehicle for the ensuing trip. Spacecraft are no different. We are sure that the captain of the *Columbia* went through his checklist to assure a safe journey on its flight from and its return to Earth (Figure 1.1). However, something went wrong or someone did not do their job. Our purpose in this chapter will be to offer some resolve, using the Columbia Accident Investigation Board's report on the disaster to emphasise the value of following processes and of not taking systematic short cuts for any reason.

Figure 1.4   Columbia at berth for check out while another Shuttle goes into orbit [Source: NASA]

In today's increasingly complex world, more complicated processes are necessary to accomplish difficult manufacturing and production operations. A good example is those used to manufacture computer chips. Services also are becoming increasingly complex, as with the Internet, credit card processing and the distribution of products and services over long distances. Today, businesses are really pushing the envelope of complexity. Systems are as complex as analysts, engineers and researchers wish to make them. And whenever anyone invents a new way to manage increased complexity, it is snapped up by a multitude of our businesses, their executives and the resultant leaders or managers. This energy and desire for more complexity cannot be terminated; it is driven by a simple concept known as the profit motive. Profit is the 'psychic' energy that drives our businesses, motivating those who are driven by this complex motive and the results it provides them.

> The risks inherent in these technical systems are heightened when they are produced and operated by complex organisations that can also break down in unanticipated ways. [5]

Increasingly complex operations become dangerous when they are run by a management structure that was created during or shortly after the industrial revolution, hundreds of years ago. The chain of command, a silo form of management structure often used today, cannot handle the difficulties created by the current complex systems. Traditional bureaucratic structures were designed to control stable operations, producing fairly simple products. Today's complex operations and increasingly complex products cannot be satisfactorily managed by the old management structures. Noticeably, the CAIB report points out that NASA uses a traditional structure to manage a complex process. Because it is totally funded by the United States Federal Government, it also takes its orders in light of the inclinations of Congress and those in power.

As this author did his research on the 2003 crisis it was interesting to find an article written in 1987 that pointed out some of the characteristics that would rear their head again on 1 February 2003. The title of the article is 'Disaster by Design and How to Avoid It'; its authors point to four organisational styles that philosophically deal with disasters: Inactive, Reactive, Proactive and Interactive philosophies. The article considers the Shuttle *Challenger* disaster and the reasons for a technological failure on that occasion. The failure of the 'O-rings' was traced to what the authors called a breakdown in NASA's chain of command. They pointed out that prior to the *Challenger* accident, there had been a long string of messages warning of the possible failure of the O-rings, but the warnings did not seem to make it through the organisation for them to be able to fix the problem [12]. The authors go on to say that while the O-rings were one of the major causes of the accident, they were not completely responsible for the tragedy that followed. The investigation showed that there was a complex web of contributing factors – technological, managerial and political – all of which needed attention if this disaster and any subsequent disasters were to be prevented in future [12].

These are some of the same issues and consequences that were discovered regarding the *Columbia* disaster. The authors of the 1987 article go on to say that the 'point at which the chain of events leading to the crisis can be broken depends upon the immediate situation, but also can be consciously chosen as a part of the overall management strategy' [12]. There is no question in the present author's mind that the safety factors involved in the NASA Space Shuttle Program were being grossly ignored and disregarded in favour of other factors.

The Columbia Accident Investigation Board (CAIB) attempted to discover the conditions that produced the tragic results of flight STS-107, and therefore is an invaluable resource for anyone studying and managing complex operations. The author highly recommends reading it. One of the Board's questions was how were the signals (of foam damage) missed? 'How could NASA have missed the signals the foam was sending?' [2].

This had been true not only during the *Columbia* flight, but in all previous Shuttle flights. What the board found was probably more than it set out to: a fundamental flaw in the organisational structure of NASA.

> The investigation revealed that in most cases the Human Space Flight program is extremely aggressive in reducing threats to safety. But ... detecting of the dangers posed by foam was impeded by 'blind spots' in NASA's safety culture. [2]

In the final report, an entire page was devoted to an analysis by Dr Edward Tufte, condemning the use of PowerPoint slides used for the engineering analysis

> At many points during its investigation, the Board was surprised to receive similar presentation slides from NASA officials in place of technical reports. The Board views the endemic use of PowerPoint briefing slides instead of technical papers as an illustration of the problematic methods of technical communication at NASA. [9]

Tufte summed up the CAIB analysis appropriately stating that 'The marked paragraph is astonishing, as members of the CAIB clearly had enough of the PowerPoint (Slides) from NASA. Serious analysis requires serious tools' [10].

It should be pointed out that Dr Tufte had performed a similar analysis on viewgraph use prior to the *Columbia* incident with respect to the *Challenger* accident of 1986, and had stated how this had contributed to the Shuttle's destruction by obscuring the relationship between temperature and the O-ring sealing. The warnings and his reports to NASA were disregarded without reason.

1.2   The *Columbia Report*: a short review

I do not wish to find fault with any person's decisions; instead I want to look at the processes in place that possibly led to the inevitable problem or problems that resulted in this disaster. I will try to highlight how some processes or the lack of them may make it inevitable that errors would occur.

To put this in perspective, we must start at a beginning. What is the mission of NASA? NASA was originally created to generate technology necessary to explore space, to take on the 'final frontier'. This has been a valid mission for the majority of NASA's lifecycle from the Mercury and Apollo missions, to the Shuttle Program. However, with the transition to the Shuttle Program the mission of NASA changed. In order to prove that the programme was a viable way to access space, NASA's mission became one of launching Shuttles on schedule and producing regular flights to a space station located in orbit. That change from its original organisational mission fundamentally changed NASA.

The heads of NASA and Congress effectively now wanted to run NASA like a business. Like most businesses, their primary goal was the R & D phase, where the new product was developed, before moving into the production phase, where in a for-profit company the product moves from experiment to a producer of cash or income. Management then shifted to a priority of reducing cost and maximising the resulting profits. This works if your product is mature, such as a car, refrigerator, aeroplane, etc. However, the Space Shuttle was still a highly experimental space-craft. NASA and Congress, respectively, must have thought that just by changing NASA's mission they could force the Shuttle into product maturity. This was considered a big mistake by many.

When a business is moving into production, management looks to reduce overhead expenses. Usually, training is reduced, except that which is needed by customers, and safety is likewise reduced, that is other than that required by OSHA. This became NASA's accepted and operational approach.

Despite periodic attempts to emphasise safety, NASA's frequent reorga-nisations in the drive to become more efficient reduced the budget for safety, sending employees conflicting messages and creating conditions more conducive to the development of a conventional bureaucracy than to the maintenance of a safety-conscious research-and-development organi-zation. Over time, a pattern of ineffective communication has resulted, leaving risks improperly defined, problems unreported and concerns unexpressed. [5]

Figure 1.5   Shuttle linked to Sky Lab [Source: NASA]

1.3   What do we think went wrong and what can be done about it?

In reviewing the operational processes that failed, questions were asked of what was bypassed and why were some of the safeguards and responsibilities neglected or not assigned. Failures often occur due to oversight of a required process. Our interests should be as to whether the oversights were due to the lack of perceived need, or were they just ignored?

NASA management was not conducting itself in a dangerous or malicious manner. They were working hard to accomplish the mission Congress had given them: regularly scheduled flights into space with the required tests and projects. The problem appears to be that a safety mentality interfered with the regular launch and project production directed by their new mission. Safety focuses on making sure everything is conducted according to the accepted processes, correct according to established protocol and then appropriate to the standard launch. Deadlines are not important to safety. Safety departments have a disdain for deadlines, and vice versa; production departments have a disdain for safety restrictions. Each makes life more difficult for the other. It may be considered that NASA saw safety as in conflict with its primary mission of regularly scheduled flights and the planned projects, in spite of the dangers cited by its safety engineers.

This conflict was pointed out by the *Challenger* investigation team and very well illustrated in the article by Mitroff and McWinney [12]. NASA's management, while not conducting itself in a malicious manner, were acting in support of what they believed were their political obligations to the programme. They didn't take the necessary action to develop preventive solutions because they had pushed the safety requirements so far down the ladder of requirements and configuration that they didn't show up on their radar [12].

Mitroff and McWinney point to four cells of preventive action that can be taken when dealing with disaster prevention. The first two cells are: (1) technical and (2) economic. In cell 1, they suggest that the available preventive actions include better detection, tighter system security and tighter internal controls, which include management, the chain of command and design changes. In cell 2, it is suggested that expert monitoring systems along with wider system-wide monitoring and mandated reviews be established [12].

The second two cells are (3) people and social and (4) organisational. In cell 3 Mitroff and McWinney suggest that detection training take place along with more social support groups and media relations training. The fourth cell deals with the organisation where there are more preventive measures established to reduce the potential of any possible disaster. The recommendations are to establish crisis management units that look for potentials (safety?) and establish a group that looks at the organisational culture for any redesigns that can be encouraged. This encouragement is believed to be a means of reducing potential future crisis and research into any potential problems [12]. All of these avenues for the reduction of potential crisis were not brought to the forefront or even examined by NASA. Instead they continued on as though nothing had happened and continued to ignore the safety issues.

To meet its primary objective, NASA began to protect itself from the Safety examinations. The Shuttle Program's complex structure erected barriers to effective communication and its safety culture no longer asked enough hard questions about risk. [Safety culture refers to an organisation's characteristics and internalised by its members – that serve to make safety the top priority.] ... The safety organisation moved from the driver's seat, in the early launches, to the rear seat, it moved to an advisory capacity. NASA management did not want safety to have veto power on launches. NASA's current philosophy for safety and mission assurance calls for centralized policy and oversight at Headquarters and decentralised execution of safety programs at the enterprise, program and project levels. Managers are subsequently given flexibility to organise safety efforts as they see fit. [3]

Safety was taken from mahogany row – or the executive suite – to the field headquarters for each launch programme. It was put under the responsibility and authority of the very management it was supposed to oversee. This placed safety in direct conflict with the launch and its fundamental requirements.

A more independent status for safety would have been a conflict of interest with NASA's primary purpose. ... No one office or person in Program Management is responsible for developing an integrated risk assessment [plan] above the sub-system level that would provide a comprehensive picture of total program risks. The net effect is that many Shuttle Program safety, quality and mission assurance roles are never clearly defined. [6]

The focus of safety was on operations at the shop floor; however, quality in producing the flight-ready Shuttle and preparation of the products and projects for each flight was the real operational objective. Safe components could be ensured and with proper planning and quality control not affect schedule. However, safety concerns prior to launch and especially after launch could cause a schedule delay – a definite conflict of interest for management. By making safety an advisory capacity, NASA could still have a safety department, but keep it from interfering with its primary mission, scheduled production and projects with a scheduled space flight.

The Board believes that although the Space Shuttle Program has effective safety practices at the 'shop floor' level, it's operational and systems safety program is flawed by its dependence on the Shuttle Program [management]. Hindered by a cumbersome organisation structure, chronic understaffing and poor management principles, the safety apparatus is not currently capable of fulfilling its mission. [4]

While not a conscious action of NASA management, pushing safety down to the programme and shop floor level effectively eliminated the safety effort and its oversight function. It had become merely an advisory group, which management could accept, reject, or ignore as it wished.

Figure 1.6   Columbia lifting into orbit [Source: NASA]

As the review Board found,

during STS-107 Flight Readiness Review, the failure to mention an outstanding technical anomaly, even if not technically a violation of NASA's
own procedures, desensitised the Shuttle Program to the dangers of foam
striking the Thermal protection system, and demonstrated just how easily
the flight preparation process can be compromised. [8]

Potential foam strikes became just a warning, not a mandate to fix the problem,
which was not seriously looked at while the crew was in space at their docking
location at the Sky Lab.

The premium placed on maintaining an operational schedule, combined
with ever-decreasing resources, gradually led Shuttle managers and engineers to miss signals of potential danger. Foam strikes on the Orbiter's
thermal protection system, no matter what the size of the debris, were
'normalised' and accepted as not being a 'safety-of-flight' risk. [8]

Because foam had struck the Shuttle during prior lift-offs, it was considered no longer a concern.

Safety is discussed here because of its published role in the *Columbia* accident. However, complex operations have many issues which can impact productivity. These are managed through several engineering approaches. One of which we shall discuss is risk analysis. Every technical or engineering organisation uses a risk assessment process to examine the likelihood of a failure. In NASA this is called 'Hazard Analysis'. NASA contracted preparation of the Shuttle for launch to a private company called United Space Alliance. It is important to look at the United Space Alliance, which is responsible for both orbiter integration and Shuttle Safety Reliability and Quality Assurance. The alliance delegates Hazard Analysis to the Boeing Company. However, for some reason, as of 2001, the Shuttle Program no longer required Boeing to conduct integrated Hazard Analyses. Instead, Boeing was instructed to now perform Hazard Analysis only at the sub-system level. In other words, Boeing analysed hazards to components and elements, but was not required to consider the Shuttle as a whole.

[Boeing] cannot effectively support the kind of 'top-down' 'Hazard Analysis' that is needed to inform managers on risk trends and identify potentially harmful interactions between systems. [6]

The business of making sure components are as safe as possible was being carried out. However, Boeing was not in charge of safety of flight for the Shuttle, and as a whole disregarded the very essence of the Shuttle Program. The interaction of all of the components which could cause an accident was still under the control of those whose primary responsibility was to prove the Shuttle was as safe as an airliner, could maintain a schedule and could conduct projects in space.

A total of 4,222 Critical Item List (CIL) items are tracked, the loss of any of which could result in the loss of the orbiter and crew.

A waiver is granted whenever a CIL component cannot be redesigned or replaced. More than 36 percent of these waivers have not been reviewed in 10 years, a sign that NASA is not aggressively monitoring changes in system risk. [7]

NASA had tried to address the CIL items. However, those that could not be addressed were redefined as non-critical as a matter of course, without any reasoning or reports provided. As budgets became tighter no further technical efforts were made to address CIL items by anyone at NASA. Risk management was to all intents and purposes whitewashed, non-existent, out of the picture for the remaining flights.

NASA told the investigating board after the *Columbia* accident that there were no safety of flight issues about which it could have done anything anyway. Because NASA could not address these issues, for budgetary or any other reasons, they decided they were not an issue and let them slide. Safety should not be ignored simply because it cannot be handled under current budgets, timelines or personnel.

You identify a safety issue and then get the resources to resolve it; this is what risk mitigation is all about. Engineering should never have agreed to this approach. The lives of the astronauts and life cycle of the equipment are of too great an importance. However, here the reverse happened. In a later chapter we will discuss the fundamental concept of requirements analysis and management; this was grossly ignored in this process.

Even after the *Columbia* launch and NASA had become aware of the foam hit, it still refused the damage assessment team's request for photos of the wing and underside of the Shuttle. Again, the thinking was more concerned with schedule than the safety of the crew and the equipment.

> When managers in the Shuttle program denied the Debris Assessment Team's (DAT) request for imagery [of the shuttle in orbit], the DAT was put in the untenable position of having to prove that a safety-of-flight issue existed without the very images [and proof] that would permit such a determination. This is precisely the opposite of how an effective safety culture would act. Organisations that deal with high-risk operations must always have a healthy fear of failure – operations must be proved safe, rather than the other way around. NASA inverted this burden of proof. [8]

Again, because NASA's mission was to produce regular flights into space and the resulting research of projects, it had to subvert and displace the safety culture that would have shown the problems that existed. The Shuttle was not yet operational for its return to earth, but no one, especially not NASA, was willing to accept this fact, dooming the crew and the Shuttle itself. As a result of this decision, it would lose all of the results and materials from its projects during this mission.

> The investigation uncovered a troubling pattern in which Shuttle Program management made erroneous assumptions about the robustness of a system based on prior success rather than on dependable engineering data and rigorous testing. [2]

The fact that the Shuttle had flown before under what were considered similar conditions was enough reason to believe it was safe. The safety culture had to prove fault, rather than require a safety analysis of the existing conditions.

Within the reorganised NASA, safety operations became dependent on production for their resources and survival. If safety became too critical, then production management could simply indicate its displeasure by reducing its budget, cutting headcount or reducing resources. This powerful message was not lost on the safety department managers.

> In reality, such a process demands a more independent status [for safety] than NASA has ever been willing to give its safety organisations, despite the recommendations of numerous outside experts over nearly two decades. [3]

Figure 1.7   Material debris for the review team to assess [Source: NASA]

Since the *Challenger* accident, outside experts have recommended to NASA that safety have its independence. However, even after the *Columbia* accident,

> it is the Board's view, shared by previous assessments, that the current safety system structure leaves the Office of Safety and Mission Assurance ill-equipped to hold a strong and central role in integrating safety functions. NASA headquarters has not effectively integrated safety efforts across its culturally and technically distinct Centers. In addition, the practice of 'buying' safety services establishes a relationship in which programs sustain the very livelihoods of the safety experts hired to oversee them. These idiosyncrasies of structure and funding preclude the safety organisation from effectively providing independent safety analysis. [4]

Safety had been effectively hobbled. It was not able to do its mission.

> Given that the entire Safety and Mission Assurance organisation depends on the Shuttle Program for resources and simultaneously lacks the independent ability to conduct detailed analyses, cost and schedule pressures can easily and unintentionally influence safety deliberations. [5]

The NASA organisation structure:

> places Shuttle safety programs in the unenviable position of having to choose between rubber-stamping engineering analyses, technical efforts, and Shuttle program decisions, or trying to carry the day during a committee meeting in which the other side almost always has more information and analytic capability. [5]

The safety managers had lost the position and weight they needed to influence production. They could not correct safety problems, only make recommendations, which could be easily ignored by production management and disregarded as unnecessary. Production management's mission was regular scheduled space flight with projects as committed, at reduced cost. They set about accomplishing their mission.

> The flight readiness process, which involves every organisation affiliated with a Shuttle mission, missed the danger signals in the history of foam loss. ... The same conclusions, repeated over time, can result in progress eventually being deemed non-problems. An extraordinary example of this phenomenon is how Shuttle Program managers assumed the foam strike on STS-107 was not a warning sign. [8]

Complex processes require a management approach different from the typical production environment assumed by NASA. In a production environment like this, cost reduction or profit enhancement works. Operations that do not focus on safety can have devastating results, such as that which occurred in the *Columbia* accident. We hope to provide a focus on the challenges of managing complex operations, as well as insight into potential approaches to address these difficult issues.

Complex organisations require a leadership style that embraces four fundamentals: (1) an applicable personal leadership capability, made up of the self; (2) people leadership characteristics that will encourage others to follow them in an acceptable manner; (3) an operational style that will provide the tools to develop effective and applicable processes and the necessary operations; and (4) an organisational style that will use the tools provided by the organisation to implement the processes into the functional culture.

These skills can best be described by a presentation that the author was privy to while working for the University of Tulsa, in April of 2010 at the commencement of the ELITE Program Class of 2010. The ideas and descriptions used were taken from an approach developed by Brian Guderian, Vice President of the Williams Exploration and Production organisation of the Williams Companies, and his staff in Tulsa, Oklahoma. The model was called the Williams Leadership Model [11], and was presented as an incentive to the graduating class in the ELITE Program in 2010 to succeed in their endeavours. While the Williams Model does not specify the components of each of the fundamentals, it was a good start and approach for evaluating and embracing management applications from an organisational viewpoint.

Subsequent chapters will add concepts and approaches that can be used with this approach. Each of the fundamental skills itemised by the Williams Model has been more effectively transposed to a new model with four leadership skills and/or concepts now used in the ELITE Leadership Program at the University of Tulsa for all the factors of an essential leadership approach. These four skills or concepts will be covered in more detail in subsequent chapters of this book.

Obviously, things were changing at NASA. It is not the author's place to question the logic used by each of the leaders within that organisation, or even those within the sub-contracting structure, but we can question the operational leadership and the people leadership functions. A huge safety question was raised by the CAIB when it assessed both the *Columbia* event and the *Challenger* incident. That question raised by the CAIB was of the position and emphasis NASA placed upon the safety review of the Shuttle mission, the crew and the Shuttle itself. An analysis by this author shows that the new requirements now placed safety as a low priority for management and operational review. There is no question about that. Congress, in the minds of the NASA administration, wanted more scheduled efficient take-offs and landings and they wanted missions conducted in a frequent and orderly manner. Without question this changed the configuration for NASA's mission orders and operations. What was not acceptable was that safety was placed so low in the order of importance and operation, or, as in many cases pointed out by the CAIB, why it was not heeded or considered a major item of discussion and interest.

Sub-contractors to NASA itemised their frustration that when the issue of safety was listed, it was often set aside and totally disregarded. So here we have two items in the overall plan being disregarded: the change in requirements and the configuration. This leads us to believe that the Shuttles' plan was devoid of the safety issue and that sub-contractors were led to believe that it was not an issue for discussion or concern, now that quality had come into focus. Only the number of missions and the ability to get the Shuttle into orbit and land it seemed to be in that arena. If the safety of the crew, the mission and the Shuttle were not an issue, then what happened to the quality of that mission and its operation? The failure of *Columbia* was certainly an issue of quality; however, there was none at all.

To illustrate this disregard for the issues of safety, one only has to read the articles released shortly after the tenth anniversary of the disaster. One of the closest associates of the crew wrote an article for the *Aviation Week and Space Technology* magazine [13]. In it he stated that the whole incident had come as a surprise to him. While he promotes the idea that the crew be honoured as heroes there is no mention of the CAIB study and the evidence that shows that NASA was neglecting its responsibility to the safety of the crew and the vehicle. While it is important for us as a nation to allocate our funds and efforts towards a viable space venue, we cannot ignore the fact that we were once such an entity, and that we disregarded safety for the mere fact of launch and study. The original configuration that included safety in its processes must be restored if the effort is to be successful in the future, no matter what kind of vehicle we choose to use [13].

Much in the following chapters returns to a lot of what we are discussing here. Processes such as requirements, configuration, planning, quality, sub-contractor management, tracking and oversight all fall into the third component (operational processes) of the ELITE Leadership Model under the concept of Process or Program Management (Chart 1.1).

These items under the ELITE Leadership Model are adapted from the Capability Maturity Model Version 1.1 (CMM) concept called Process or Program Management, and should be a concern for any leader. The other arenas in this component of the ELITE Leadership Model are Project and Systems Management, which I am sure the NASA management will feel they did well in, and Business Acumen. However, it would not be uncommon for the reader to see this author questioning that. The last is good judgment, which is obviously not one of the good points for NASA management, as noted in the CAIB Report. By putting safety into a subordinate and often non-existent position for the decisions of 'go-or-no-go' on the mission they were certainly not using clear and concise judgment that would be necessary for a quality Shuttle flight. It was evident that NASA had learnt nothing from the *Challenger* disaster many years earlier. Events such as these are supposed to be learning experiences, yet safety had been excluded because someone saw the changing requirements placed by Congress as being more important, both informational and financially.

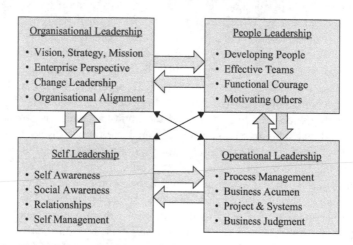

Chart 1.1   The ELITE Leadership Model [Source: Reprinted with permission of the University of Tulsa, ELITE Program [14]]. Note: See Appendix section for full page image

Of course there are a lot of other questions that need to be answered. But for now let us be content that NASA may have caused some major problems with its decision to reduce the effect of safety concerns. Let us also realise that many of you will see leadership very differently when you have finished this book. Leadership today requires more than a seat of the pants assessment of a situation under any condition. Any situation is almost mathematical or exponential; that is to say, quantum physics is alive and well in our current environment. One decision can ripple throughout each of the operations and functions in today's complex organisations, with devastating effect. Without a plan and a real set of working key processes in hand the results may be as devastating as the *Columbia* disaster was for NASA.

Questions for the reader

1. Can you provide some reasons why you think high technologies as we see them today are so difficult to manage?
2. What would you have suggested to NASA that would have improved their safety operations, reduced the incidents and provided an answer to their needs

in providing for more successful launches? Under the existing current conditions, as cited in the chapter, do you think your suggestions would have had any effect?

3. What role do you feel organisational leadership, as cited in the ELITE Leadership Model, plays in the overall NASA project plans and how would these key elements help in their ability to provide for success?

4. What role do you feel operational leadership, as cited in the ELITE Leadership Model, plays in the overall NASA project plans and how would these key elements help in their ability to provide for success?

5. What are your feelings or opinions about the comments made by the researcher regarding the use of PowerPoint slides by the staff working for NASA? Do you personally feel this had an impact on the disaster or the problems in operations?

6. When one considers that operational leadership is required to guarantee success in NASA and the Space Shuttle Program, where do you feel they may have failed?

7. Can you discern from previous reports whether there was evidence that damage was being done to the shuttle in their missions, and not only to this one (STS-112)?

8. NASA has stated that the existing budget would not allow them to deal with the safety issue. First, do you really believe this statement and what would you suggest to them as an Engineer on the Shuttle project, regarding this situation?

9. What role did the various projects have in the decisions that NASA made about the Shuttle launches?

10. Describe how you would deal with the *Columbia*'s issues using the ELITE Leadership Model? Use each of the four segments (Components).

11. Can the ELITE Leadership Model provide you with a solution to the NASA situation or does it confuse you more? Can you tell us why you've come to this conclusion?

12. Why do you think the NASA leadership decided to ignore safety and the potential of a 'G0 No GO' situation in each of their launches?

13. With all the data that was collected after the *Challenger* disaster, what do you think was the reasons they didn't see the *Columbia* disaster as a potential?

14. With all of the data staring the NASA management in the face, why do you think they were reluctant to allow the crew and the Sky Lab to do its due diligence on checking out the damage to the wing?

15. What specific items in the operational leadership side of the ELITE Leadership Model do you think might have had a more effective result if it had been used by NASA?

16. How well did NASA do on all the items in the ELITE Leadership Model when dealing with the issues at hand and the issue of the foam damage on previous shuttle flights? Is there something that you would have done to change that?

References

1. Columbia Accident Investigation Board. *Columbia, Report Volume 1.* Washington, DC: National Aeronautics and Space Administration; 2003, p. 12
2. Columbia Accident Investigation Board. *Columbia, Report Volume 1.* Washington, DC: National Aeronautics and Space Administration; 2003, p. 184
3. Columbia Accident Investigation Board. *Columbia, Report Volume 1.* Washington, DC: National Aeronautics and Space Administration; 2003, p. 185
4. Columbia Accident Investigation Board. *Columbia, Report Volume 1.* Washington, DC: National Aeronautics and Space Administration; 2003, p. 186
5. Columbia Accident Investigation Board. *Columbia, Report Volume 1.* Washington, DC: National Aeronautics and Space Administration; 2003, p. 187
6. Columbia Accident Investigation Board. *Columbia, Report Volume 1.* Washington, DC: National Aeronautics and Space Administration; 2003, p. 188
7. Columbia Accident Investigation Board. *Columbia, Report Volume 1.* Washington, DC: National Aeronautics and Space Administration; 2003, p. 189
8. Columbia Accident Investigation Board. *Columbia, Report Volume 1.* Washington, DC: National Aeronautics and Space Administration; 2003, p. 190
9. Columbia Accident Investigation Board. *Columbia, Report Volume 1.* Washington, DC: National Aeronautics and Space Administration; 2003, p. 191
10. Tufte, E. *Columbia Accident Investigation Board: The Boeing PowerPoint Slide* [online]. 2003. Available from http://www.edwardtufte.com/bboard/q-and-a-fetch-msg?msg_id=0001yB [Accessed on 30 May 2013]
11. Guderian, B. *Leadership Development and the Role of Continuing Education.* Presented to graduating class of the ELITE Program; Tulsa, OK, 2010, p. 2
12. Mitroff, I.I., McWinney, W. 'Disaster by Design and How to Avoid It', *Training Magazine.* Aug 1987, pp. 33–8
13. Ross, J. 'Space Needs New National Consensus'. *Aviation Week and Space Technology.* Jan 2013, p. 50
14. Patricia Hall, The ELITE Leadership Model, The University of Tulsa, 2013

Chapter 2

The importance of processes

Figure 2.1   Boeing P-6 ready for flight

Processes really are important. They are the 'what' that needs to be accomplished to successfully complete a job relating to the project, production or service effectively and efficiently. But then again, it may be thought that for detailing everything what you have to do is a simplistic and unnecessary exercise; it may also be considered unnatural and too matter-of-fact. If we were pessimistic, we might say, 'Any mature or sane person would know that this is the way this job should be done'. However, the most effective engineers, especially those in NASA, now know that ignoring a process is not always the wise choice! Many processes are developed through the painstaking efforts of many experienced engineers or practitioners (often a systems engineer) who have worked with the same function, task or activity, studied it and attempted to make it more practical, meaningful and productive. In most business operations, the results are defined, but in too many cases the 'how' required to achieve the results of the 'what' are loosely defined and left to the employee's initiative. Ignoring what we know to be true, such as the 'how', can lead to chaos and gross errors of monumental proportion, as seen in the previous *Columbia* case study.

A process in engineering can be best described as a sequence of steps performed for a specific purpose. It can also be a set of interrelated activities that together transform inputs into outputs; where a set of activities, methods, practices and transformations that individuals use to develop and maintain a product are applied in their most realistic sense. The associated products, such as project plans, designs, documents, test cases and work packages result from the inputs provided by the requirements through the stakeholder to produce the product, project, service or output.

Therefore, the process itself has several elements. First are the specific tasks, steps or activities that must be done. This is where the task can be assigned to a person, stakeholder or employee to execute with the responsibility and authority to control and complete the required activity. Here the task may have several sub-tasks as a part of its activity. The methods and tools used to execute the activity are the second element. These set specific guidance and direction to the activity where the method or tool criteria prescribe a systematic, repeatable technique to support the tasks further. Tools might include the scheduling methods used such as a Gantt, or PERT charts, and other automated tools like *Microsoft Project*. The last element of the process is the work product, project or service itself that results from the work packages identified. This is where the data is collected or the product is finished for subsequent activities or final assembly of the unit.

This is why you will hear experienced systems engineers insisting on knowing what the requirements are for a particular product, project or service operation. The same engineers might also ask what the 'baseline' is for the current process, thereby wanting to know what determined the existing tasks or functions and the activities that incorporate this process and how they are put together. With this knowledge the engineer can see what is intended, what requirements analysis was conducted, what configuration might be required, what tools and methods are being used to function and the work packages or outputs needed or intended.

A knowledgeable chief engineer, systems engineer or engineering project or programme manager will instinctively know that there are six key processes that they will have to work with from inception of the product to its eventual completion as a work product or package, and then to the first article in its production. Those six key processes are:

1. requirements management,
2. configuration management,
3. subcontract management,
4. quality assurance,
5. project planning and
6. tracking and oversight.

Without these key operational processes in place and functioning, the project, programme or service will not have repeatability in terms of its output and final results. In other words, every time that the specific processes, functions or services are conducted as a work package, the product will be different in outcome due to the lack of consistency in its function, or it will be non-existent because of non-application of these key processes. These six key processes are commonly known as the key engineering operational processes and fit well with the ELITE Leadership Model as the skills required for an effective leader. These are specific controls that flow from the leader as applicable skills that are developed overtime and from the experience gained in working with or doing the key processes of project or programme management. They help reduce risk and aid the company's definition of the projects, product or service operation or function.

2.1 A brief history of process engineering

Before setting out definitions for the six operational and most important key processes, it should be of interest to the reader to know where this process knowledge has come from and what spurred research into what we now call process engineering. The venture into the actual development of processes and their applications in engineering began in the arena of software engineering. Writing 'correct' processes for software development was recognised early on as the major challenge affecting the computer revolution in the early 1960s. A breakthrough came in 1969, when Professor Edsger W. Dijsktra wrote a letter to the Journal of the Association for Computing Machinery (ACM) suggesting that the GOTO statement was a major reason for unintelligible, unreliable ('buggy') code, as referenced by Humphrey [1].

In the software engineering world, emphasis was immediately placed on developing more powerful, yet easier to use programming languages. After Dijsktra's paper, programming languages began to stress Computer Assisted Software Engineering (CASE) structures and run-time error checking as tools for fixing the problems. Frustrated by a profusion of languages, the Department of Defense (DoD) sponsored a major project initiative with several major engineering universities and the National Labs to develop a standard programming language that might reduce the confusion and incorporate the program instructions into the actual programming process. The result was the Ada programming language, used successfully by the US Military.

On the methodology front, Dijsktra's paper triggered the onset of top-down, 'structured' programming (championed by IBM's Harlan Mills). Yet after another 15 years, programming was still more of an art form than an engineering discipline. To address the importance of the need for a true software engineering discipline the DoD now funded the Carnegie Mellon University (CMU) to develop a software engineering 'best practice'. This resulted in CMU and its constituents creating the Software Engineering Institute (SEI); the resulting methodology became the Capability Maturity Model (CMM), which has been in use since the early 1990s [1].

The SEI was the first to produce a framework document for software development in early 1993. This document later became the 'software' CMM Version 1.1 (Chart 2.1) [3], which focused on developing software projects that would be

	Level	Focus	Key Process Areas	Result
Improve	**Optimising** 5	Continuous Improvement	Process Change Management Technology Change Management Defect Prevention	Productivity & Quality
Control	**Managed** 4	Product and Process Quality	Quality Management Quantitative Process Management	
Defined	**Defined** 3	Engineering Process	Organisation Process Focus Organisation Process Definition Peer Reviews Training Programme Inter-group Coordination Product Engineering Integrated Management	
	Repeatable 2	Project Management	Requirements Management Project Planning Project Tracking & Oversight Subcontract Management Quality Assurance Configuration Management	R I S K
	Initial 1	Heroes		

Improvement initiatives must increase market share and/or profitability in order to have business value!

Version 1.1 C-SEI/CMU

Chart 2.1   A Process Management Model … the Capability Maturity Model [Source: Image adapted with permission from Zubrow, D. 'Putting "M" in the Model: Measurement in CMMI', Carnegie Mellon University, 2007]. Note: See Appendix section for full page image

efficient due to their reduced flaws [1]. Much of this was based on work done at IBM by Watts Humphrey and others. Since then, other CMMs have been produced to address other aspects of the engineering and product development lifecycle. Most importantly for software, the SEI has since developed two models by combining the other lifecycle models (specifically systems) with the software CMM to become the Capability Maturity Model Integrated (CMMI) and the CMM Staged. This book will emphasise the initial model (CMM 1.1) developed by Watts Humphrey at CMU for the Software Productivity Institute (SPI). The author believes that the original CMM 1.1 has, through its use in several organisations in the United States, proven to be the foundation for effective engineering processes overall; see Humphrey [2].

The resulting six key processes for the second stage of the original CMM were generated by the framework of the CMM and the SEI to focus on repeatability and to emphasise a foundation for the overall engineering process. Thanks to the work of Humphrey at IBM and others who have since recognised the importance of repeatability to the engineering discipline, it was only a matter of time before many systems engineers began to use, establish and institute the key processes into their engineering operations for standard operating procedures. For the sound purposes of engineering and the operational skills of that field, the present author has retained CMM Version 1.1 as the operational model for all engineering fields in reducing risk and increasing definition for the overall project, product or service.

The following five levels of capability maturity used in this chapter come from CMM 1.1:

1. initial,
2. repeatable,
3. defined,
4. managed and
5. optimised.

The author will attempt to focus on those key processes, which provide the field of project and programme engineering and project management leadership with the greatest return on investment. All of the writings in the SEI on the five levels or stages repeat the importance of the second level. This level is consistently called the 'project management' or 'programme management' stage in relation to CMM 1.1, although current SEI publications describe it as the repeatable level [2]. In engineering we recognise that repeatability is a major function of our work and development of the products that result from our projects through the management of process and through the leadership role. This can only be referred to as the programme or project management role. Keep in mind that there are five levels of capability maturity and six key processes at the second level of the CMM. These will be used throughout this book.

2.1.1   Applying the principles of process engineering

When you analyse the project management (or repeatable) level of CMM 1.1, it can be seen that it truly emphasises the fundamentals or requirements of engineering project management. The requirements process is a fundamental starting element for any engineering project, so for that reason we have chosen to follow the leadership of the SEI and proclaim that the six key processes in the repeatable level are indeed the fundamentals of (and focus for) engineering project or programme management for operational leadership (an element cited in the ELITE Leadership Model). The purpose of requirements management is to establish a common understanding between the customer and the project executors so that the inherent needs of the project will be addressed. This involves establishing and maintaining an agreement between all the parties involved regarding the specific project or system needs that covers both the technical and non-technical delivery items and work packages that will satisfy these requirements. The agreements drafted form the basis for estimating, planning, performing and tracking the engineering project or product throughout its lifecycle. The goals of the requirements management process are to control the system needs in order to establish a baseline for the engineering project and provide management with a tool to use in assessing the progress of that project, programme or service. These requirements in turn will be consistently used in the engineering project's plans, products, configuration and activities whether it is a product, service or project piece of a larger operation.

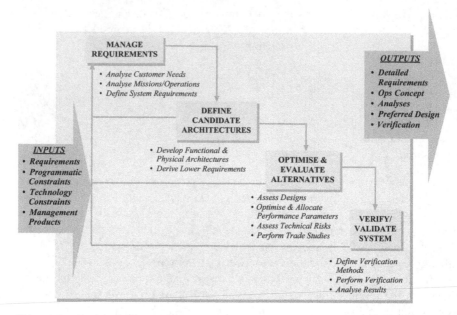

Chart 2.2   Technical Requirements Process. Note: See Appendix section for full page image

Note that in Chart 2.2, the Key is the analysis of the customer, the mission and the system requirements. Many go wrong in not defining the verification methods that will be used, performing that verification and, through the analysis of the results, verifying that the requirements are really being met. It is not that the other steps are not important, but the needs must be analysed and verified to make an effective project, programme or service available to the customer. This is where the importance of the planning process must be imposed. Part of the planning process is of course the tracking plan and the chosen oversight that will be put into place. These items will be described and their use explained later as a part of the overall capability maturity. The architecture provides the configuration plan and the quality assurance.

The activities, functions, tasks and processes for a product most often originate from an initial proposal or idea, developed to justify the need for that product. This is either done for the company by their product development organisation or developed by the responsible parties in response to a customer's desire or request for a solution to some problem operation or concern. All the activities and tasks to produce this product are determined so that cost and budgets can be estimated and determined to justify the production and to show that a profit can be made from the proposed venture. To establish that cost, all aspects of the requirements, the product development and the manufacture are examined through the use of a 'work breakdown structure' or WBS. The WBS is the first established document and results in the statement of work (SOW). It is the documented foundation for the product and should always be considered in this manner. In his research and his experiences with industry, the author has found that this is not always the case. For example, where has the WBS been hidden in the careful consideration of requirements or operational needs for the *Columbia* or *Challenger* disasters, and where is NASA's resulting resolve for solutions that would have alleviated the resulting disasters or catastrophes. Maybe it was there, but was never mentioned by the Committee?

Chart 2.3 illustrates how the WBS fits into the overall product planning process. If the WBS is not part of your current process, the author suggests that you assess this concept and the 'fit' of the processes suggested in Chart 2.3 to enable a more appropriate approach to completion of the necessary guides, policy and directives needed to support effective management of the production of products in your organisation. The WBS, as a piece of the overall foundation document of requirements, establishes the flow of work in a company and also allows management to review its policies, procedures, methods and tools. The WBS also allows for an efficient writing of the SOW. It enables all involved to see the origins of the concept, where they have been and where they want to go. In addition, it allows for corrective action, change management, planning, project tracking and the configuration that all together manage an effective and productive project and organisation. On the short side, it reduces the chaos and indecision that may be in place about the product as it reaches its development and production stages. This use of the WBS is a part of the ELITE Leadership Model that fits into the category of process management, judgment and business acumen, for without the project or programme experience to know how things work and fit together, a person would be clueless about developing the WBS. Project and systems management principles also fit well at this step. Now we need input from the organisational side, such as the product's operational definition. This input includes the company vision, strategy and mission and the enterprise perspective that incorporates the product into its activities.

Before an organisation starts to develop the WBS, it must understand the operational definition of this product. That is, what is this product, what is it expected to do, how will it work and what can the customer expect from it? This operational definition establishes the first steps towards understanding what it is that the company is trying to do, what it wants to build, service, manufacture, or establish, and the need for preliminary requirements that will be the beginning and the criteria upon which the rest of this product will be built. The operational definition sets the stage for the requirements, the WBS, the configuration and many more factors that will be worked into the overall product lifecycle. The operational definition also encompasses the use of the company's rules (vision, strategy, mission and goals) and planning process, and aids in the development of the scope and deliverables expected of the product [7].

Now you get to work on the step one of the WBS. In the pre-planning stage, the operational definition is shared with all those who are working on this project and the development of the product. The process of concurrent engineering is also suggested at this stage. This is called many things in various organisations, such as integrated product Development (IPD) and, where teams are established, IPTs (or

IPDTs) for integrated product development teams. The importance of the systems engineer as part of the concurrent team can also not be emphasised too highly. They are integral to the team in the concurrent process because they maintain the focus on the requirements, SOW, risk mitigation, scope and deliverables that support the initial operational definition. Systems engineers help the team look at the risks involved and their overall analysis, and develop the risk assessment and mitigation plans necessary to assure a viable product, a profitable outcome and a meaningful result for the company through the deliverables and work packages.

Requirements analysis is a specialty of a systems engineering organisation. They are best suited to assist the IPDT through the traps and hoops of the real requirements. The project or programme team has to analyse the needs of the customer and how to best meet those needs, and put in place all the mechanisms that will provide a solution. Product projects or programmes must include this thorough analysis if they are to function properly, with each solution having sufficient risk assessment and mitigation resolve to meet its needs. With a full knowledge of the deliverables and work packages the configuration can now be developed and shared with the players from the team. This gives each member of the team the opportunity to assess their involvement and any problems that might be inherent with the concept. The resulting requirements should then be a full consensus of agreement from all team members and a suggested configuration by all as to the deliverables and work package arrangement for operations.

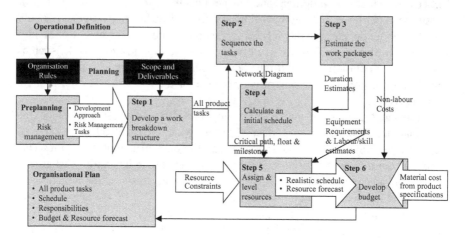

Chart 2.3   Developing the Work Breakdown Structure (WBS): Decision Making [Source: Reprinted with permission from Verzuh, E. The Fast Forward MBA. 2nd edn. New York: John Wiley and Sons; 2005]. Note: See Appendix section for full page image

The systems engineering lifecycle has seven developmental processes. It begins with the **user requirements** and then reviews the **system requirements**. This will generally provide input for changes to the overall requirements. Once the changes are reviewed, the **architectural design** is made and then reviewed against the requirements. Using the refined design the **components are developed** and reviewed against the new design parameters. After further review the components are **tested** and **integrated** into the system where they are verified as being usable and correct. Again changes are assessed and the overall system is tested for **acceptance and installation**. If the acceptance tests are verified the product will be forwarded to **operations and maintenance** where again it is reviewed and changes suggested if needed. These seven steps are required to ensure that the product, project or service is a valid function [8].

2.2   Stop 'fixing' that broken process for the umpteenth time

Processes get fixed over and over again, but in many cases with actual change to the end result. No process should go through a fix without having a thorough review by an appropriately charged and knowledgeable change control group. This group must have the authority to institute the changes if acceptable, meaningful and effective. To factor in the need, the workers, managers and affected personnel who work with the process must see change as an important function of their job. Therefore change must be an accepted process and condition in the company. Policy needs to be in place that encourages meaningful change, improvement and cost savings. People must be incentivised to encourage change and they must receive rewards rather than just be seen as maintaining the existing process. In addition, changes should not be made without them having gone to this suggested control group, and they should not be recorded as having a positive end result without scrutiny by the group. The recording process should be part of the overall change element so that it is documented as a process. The result of the assessment should be issued to all who use it. This is why it is so important for these reward elements to be involved in recommending change, supporting the change, documenting it and then using it in its new state to provide greater return to the company and the product.

I mention the above because it is why key process number two is so important. Configuration management has encouraged us to put things into their accepted architecture – that structure is based on the requirements as currently accepted and baselined known by all and from which all operate. Configuration management establishes the guidelines and structure by controlling the baselines and maintains the integrity of the product over time through an established process. Configuration management provides the system that controls the changes, maintains the integrity of the product and traceability of changes throughout the product's lifecycle. It is suggested here that the configuration system establish a baseline data operation or library collection that functions with the controls on the overall product configuration, the changes and the auditing processes that are necessary for all involved, and where the baseline and any changes to it can be recorded and shared with anyone dealing with the product or project.

The goals of configuration management are to have planned processes that identify the product, control the changes and inform all affected groups and individuals of the status and content of the existing baselines. This is the control that keeps outside change from happening until all are informed via the final approval from the change control organisation or group. Everyone will then be working from the same baseline model towards developing of the final product. This prevents someone from having a brilliant idea at some point and changing the configuration halfway through the project, leaving others to wonder what happened when the final outcome didn't look like what everyone expected or what was really approved. This type of event can have a devastating effect on any outsourced work packages or unsuspecting subcontractor who has taken on a product development phase for the company.

The questions that must be asked as the company begins the project or the development of the product are based on a requirements list. Are the requirements baselined and under some sort of configuration control? As the engineering requirements change, are adjustments made to the engineering plans and the associated work packages or products? Does the project follow a written organisation policy for managing the system requirements? Are measurements taken to determine the status of the proposed changes to the baselined requirements? Is there a documented engineering configuration management plan? Are the work packages (products) under configuration management control? Does the company follow a documented procedure to control changes to the work packages? Are standard reports on baselines and change requests (e.g. Change Control Board Minutes) distributed to the affected groups and individuals? Using these data points as guidelines will help the Project or Product Engineering Leader execute the appropriate steps and processes throughout the entire project (see Chart 2.4).

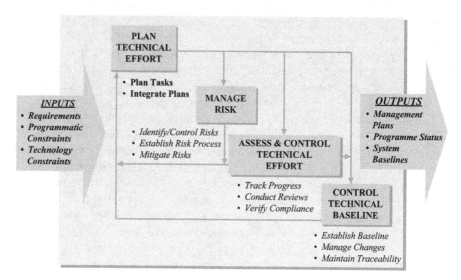

Chart 2.4   A Systems Management Process. Note: See Appendix section for full page image

Broken processes exist because there either is no configuration to begin with, or there are no controls on the existing product integrity processes that require change management, or there was no consideration for the project at all within the configuration system. Again, without these controls and appropriate answers to the questions above, the result can destroy a good product, project or service agreement. The change control board or group (CCB or CCG) is an effective part of the third leadership component; this is characterised as an organisational skill. Change leadership is a strong skill that can only be developed over time through experience. It is developed as a result of seeing and experiencing the failures of the past due to a lack of requirements or configuration management. With this skill, one often allows the company to alleviate embarrassing situations and events. Change leadership uses the developed components of the operational leadership skills that function as business acumen and judgment (experience based) using the skill of configuration management and analysis. This capability comes from experience and time on task and cannot be written off as a known experience; it must be acquired.

There are many systems in which an industry reviews its processes and makes suggestions to change and improve their operation. Unless this procedure is established to look at the entire relationship of the product baseline to the organisation, a simple process change may have little or no effect on improvement or may result in a total change in direction for the product. This is also a good example of what was happening in NASA as the Shuttle Program matured. For example, a fix mechanism such as value stream analysis is often applied to one sub-process of an overall organisational operation. If the company does not assess how the identified change will affect the overall operation (i.e. the whole process), a simple change may have little or no effect and may even increase the problems faced by the company. A good example is the term used often in industry of looking for 'low hanging fruit', where the fruit often turn out to be sub-processes that are part of a

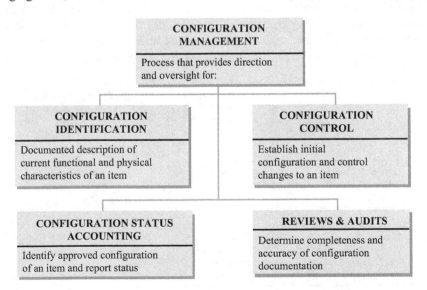

Chart 2.5   What is Configuration Management? Note: See Appendix section for full page image

bigger process. When the sub-process is changed, it changes the configuration to such an extent that it may throw the entire process out of whack, costing the company more money and resulting in errors that might not be able to be fixed, such as the conditions in the Space Shuttle's operational processes. This is why the author supports the establishment of CCBs and the establishment of a policy for configuration management where people who are knowledgeable and experienced about the company and its processes can look at the overall effect, and can encourage the appropriate changes necessary. Along with the necessary institutionalisation of those changes throughout the company, this will provide for effective results to overall operations. This step can also reduce the effect a minor change would have to an overall system and would bring about a broader change that is really needed.

CCBs can be an effective and efficient means of baseline control through the configuration management system if they are a key process for effective product or project management with the desired result of a positive nature (see Chart 2.6). The *Challenger* and *Columbia* incidents would never have happened with an effective and efficient review of the safety issues involved in a change of process that involved a CCB as opposed to a simple management decision.

The third key process in the project management stream is one of the most overlooked of all. Subcontractor management can make or break a project. Today we are hiring more subcontractors and outsourcing more operations than ever before. With this reality, many companies are still not effectively looking at the requirements that make subcontracting a workable process. The leadership and management of the subcontractor cannot simply be a search for the best proposal or lowest cost from a prospective contractor. This approach is most often used because the finance or accounting organisation seems to be the key processor of the subcontractor's proposal. While cost and schedule are important, good engineering and systems analysis must also be criteria. This includes the risk assessment by the systems engineers of using certain contractors and their manufacturing methods, past history for success and product reliability based on good engineering analysis

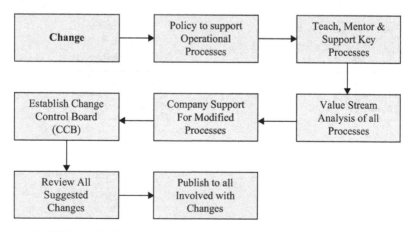

Chart 2.6   Change, Policy & The CCB. Note: See Appendix section for full page image

and review. Again, we see that the systems engineer can be your best friend in this assessment.

Choosing the best qualified personnel for the various roles is of paramount importance. It can only be done if the project managers have done their homework to focus on the exact tasks, activities and operations they expect to be carried out. Incorporated in that assessment needs to be the knowledge, skills and abilities expected from that subcontractor, as well as the processes, methods and tools they will be using. These factors, along with their past history of producing the expected work product for another client, will establish the best qualified. The factors that determine the best qualified should not be trumped by the cheapest or the best constructed proposal. Someone who can write a great proposal and puts the right words in place may not be the person who is best qualified to do the job.

The aim of subcontractor management should be to select the most qualified contractor and finalise in writing the commitments by both parties that follow the processes established at the host or prime company. These commitments will follow the requirements and configuration management processes established by the prime as necessary procedures for the subcontractor to follow in producing the required parts or services. Quality should also be a big component of the contract with the subcontractor. The subcontractor should expect the host company to track and review their performance based on the documented contract, using a collective focus on the non-technical as well as technical requirements to which they must adhere.

Both parties should commit to adhere to a communications requirement that maintains ongoing interchanges concerning the commitments, corrective actions and changes to requirements and configuration as they are approved. This type of focus will ensure that the results from the subcontractor will be more predictable, the assembly of the end product will be more precise and the relationship between the subcontractor and the requestor will be more respected and accepted by all parties. Again, the questions to ask are: Is there a documented procedure for selecting qualified and reputable subcontractors? Are periodic technical interchanges held with the sub-contractors? Are the results and performance of the subcontractor tracked against their agreed to commitments? Are the activities reviewed with the product leader on both a periodic and event-driven basis? Are there quality and precision points in place for the subcontractor to follow? Keeping these points in mind as part of the agreement, and then following them, provides for a most effective result between the prime company and the sub-contractor.

The real issue is that the leader and their staff are managing according to the configuration established for the event or activity. Whether it is the company or the IPDT organisation, the same consistency needs to be supported throughout the operation and according to the agreed-to stipulations.

2.3 Business process re-engineering only works when applied and supported

Business process re-engineering (BPR) has often been touted as the answer to all the ills that a company can experience. Books have been written to demonstrate how the process is done and why it should be applied. During the 1990s, BPR was embraced by many companies as the best approach to improve their operations. However, despite the emphasis placed on this concept, there has been very little improvement in many of the organisations who have claimed to have used the technique. Usually, there is a lack of attention to the appropriate preparation or follow through by top-level management to implement BPR. As a result, projects flop. As so often, a lack of 'walking the talk' or top-level management support for the process leaves workers feeling that things are being done to them, rather than improving the processes or operations for them and the company as a whole [4].

In order to focus on 'walking the talk', we need to introduce the fourth key process in project management. This is quality assurance (QA). This key process gives a company the ability to verify it is doing what it says it is doing or will do; that is, to 'walk the talk'. QA provides the company with all the processes and methods to review and audit the tasks, functions and activities to verify they comply with the stated procedures and standards. To be sure that the processes will be used, a QA group should be involved in the early stages of the project to help establish the plans, standards and procedures that will add value and satisfy the constraints of the project or product following the organisation's policies. Again, the company here needs another policy to establish such a group and to give it the teeth it needs to function effectively. The author realises that this is a hard issue to swallow, but without an appropriately designated and authorised QA group the whole idea of 'walking the talk' is a sham. From its inception it is the responsibility of the QA group to verify that the processes are being used and that they appropriately fit the needs of the project. In some cases the group might be the first to point out that the tasks, functions or activities as originally designed might not be the best available, and would be actively getting them changed through the CCB.

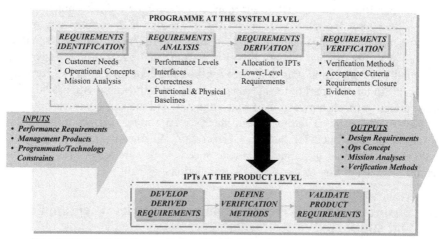

System Requirements Satisfy Customer Needs and Maximise System Cost.

Chart 2.7   Manage Requirements. Note: See Appendix section for full page image

The goals of the QA group are to have planned activities to assure adherence of the project activities and products to the applicable standards, procedures and requirements, and to objectively verify the appropriate actions. Additionally, all affected groups would be informed of the group's activities and results, and of any corrective action that was required. If issues are not resolved, senior management would be informed for them to conduct the appropriate action. QA Should not be the policeman who verifies appropriate action or punishes offenders. It is the member of the IPDT that helps workers and staff correctly execute their job according to the correct procedure and standards. QA helps employees, leaders and managers make the changes necessary when a policy, process or procedure is hindering effective product operations. It provides employees and the system with the resources to support their arguments for change when it is appropriate and necessary.

The QA group also supports employees when a subcontractor is not meeting their agreed-to commitments or requirements. Product quality is highly dependent on the processes and their appropriate execution. Repeatability, remember? If a process or procedure is hampering the correct and efficient execution of a project, then the QA group is expected to work with the appropriate people to re-engineer the process and improve the activity to meet the requirements more efficiently and effectively. This is a key area, and the following questions should be asked by management or the leadership when getting the QA group involved: Are there a documented engineering QA plan and policy that include the activities for process evaluation? Is there objective evidence of engineering QA activities? Have the results of the QA reviews and audits been provided to the affected groups and individuals? Does the project/product follow an operational and organisational policy for implementing QA? Are the engineering QA group's activities reviewed with senior management on a periodic basis? Answering these questions appropriately provides the project or product leader with good feedback on the quality issues and senior management's interest in them.

The fifth key process is project planning, the sixth is tracking and oversight. As shown previously, knowing the project's requirements, configuration and subcontractors is not enough. The plans must be reasonable so as to perform their function and meet the requirements and configuration of the project, product or service. Estimates are developed for the work to be performed that will allow the project or product manager to establish the necessary commitments and define the plan to perform as expected. As stated earlier, everything starts with an SOW based on the WBS. The planning process will include the steps shown in Chart 2.3 for estimates, size of the project, resources needed, schedule, risks and appropriate mitigations, along with the necessary commitments. This plan format provides a basis for effective performance, and it addresses the commitments and activities required. It is worth restating the importance of informing all affected members of the project team of how the leader expects the team to be aware of the plan, its estimates, activities and commitments, and how they must be able to relate to these activities as the means by which they will track and oversee the projects' progress to completion.

The tracking and oversight of the project is based on the data generated from the plan. The documented plan is the basis for verifying the results, adjusting the schedule where necessary and taking any corrective actions required. It is the project management leaders and their designated specialists who track the performance against the plan. When corrective action is taken, that action is managed to closure for the desired results, especially when deviations from the plan are found. As changes to the original plan are required, the project manager works the configuration changes with the QA group to verify the rationale and present the requests to the CCB for adjustment or agreement. Key questions that help the leader guide the project, product or service through the hoops are as follows: What are the actual schedule, size and cost compared to the original estimates? Is corrective action taken when actual results deviate significantly from the plan? Is a policy written for the organisation that enacts the tracking and oversight activities? Are changes to the engineering activities agreed to by all the affected parties? Are the performance results, open issues, risks and action items reviewed with senior management on a periodic basis?

2.4   Taking the time to plan and fix processes

Time needs to be set aside to plan out a BPR event and to establish the actual process through a training effort that includes all the workers, stakeholders and employees involved. It is always possible to change a process and train the workers to do it differently. However, if you don't take the time to instruct and emphasise through training the proper way something now needs to be done the process will not get implemented by the stakeholder in an appropriate way or in a timely fashion. In addition, if top-level management is not 'walking the talk' and demonstrating that the process needs to be done the way it has been revised or newly implemented, then the changes will not happen. Employees need to know that the changes are important to the company as well as the leaders and top-level management, and that the changes are supported for the good of all the participants – that management not only supports the change, it emphasises it.

This is why it is important for all management levels to be upfront and clear in their explanations to employees and stakeholders. Many employees do not know what the management focus is in the company, so this needs to be made clear through some form of informative mechanism. For example, most managers know intuitively that in today's business climate, customer satisfaction is one of the prime objectives. This goes hand in hand with a focus on positive revenues and sales objectives, with the unstated goal of significant returns to shareholders. However, if management does not share these essential goals in its discussions with employees, knowledge of these objectives will never be a reality in the company or on behalf of the employees' actions for the company. The importance of satisfying the customer must be communicated whether the company is a service producer or a product manufacturer. This is part of 'walking the talk' [4].

Management uses four forms of concepts when working through processes:

1.   philosophical,
2.   humanitarian,

3. logical and
4. technological.

Unfortunately workers, stakeholders and shareholders outside the company hierarchy have very little knowledge of these concepts and have no way of asking the questions that would help them to understand. The material in this chapter is intended to provide that understanding and the concepts that management must communicate. The four concepts are used to consider the requirements of the three stakeholders in the business: the shareholder, the customer and the employee. Philosophically the manager wants to support the three stakeholders with a positive approach that will give each the most positive results and make the manager feel good about what they have done with the product; that is, something good for the customer and shareholder/employee alike. The manager wants to be profit-oriented, customer-driven, partner-assisted, employee-centred, and environmentally and ethically conscious. On the humanitarian side, the manager realises the need to convince the company's top management and its employees and partners and the need to continuously reinforce its objectives to each of those parties. The logical side focuses on the planning, organisation, company direction, control and assurance. The technological side looks for the best approaches to get the job done, exercising BPR, lean management, value stream analysis, continuous improvements and other methods as necessary.

2.5 Are there stakeholders in each action plan for the processes?

Who are the stakeholders? They are the customer, the investors or shareholders, the employees and all levels of leadership and management. Once the focus is flow of the product, profit and improvement, many people who would have taken credit are no longer involved. Instead, the immediate stakeholders may be those who provide the direct input, throughput and output essential to productivity. This comes about if you accept the model presented in Chart 2.8. A company's management will consider the four forms of concepts if you accept that there are three major stakeholders involved in Chart 2.8: the customer, the shareholder and the employee.

How are the four concepts involved? Management's philosophy may be profit-oriented, customer-driven, employee-centred, partner-oriented or environmentally or ethically oriented. What do we need to do about their involvement to make sure they have a feeling of ownership in the processes and their improvement?

On the basis of Chart 2.8, we can see that the managers' viewpoint is not one that we see often or might be the perception of the average company stakeholder.

The manager or leader is beset with large considerations right from the very beginning of their involvement in the company's operations. With all of the philosophical considerations – especially pleasing the stakeholders – the leader must be concerned with those stakeholders who have diverse interests and random requirements of the company. In order to establish a positive company climate for all the stakeholders, the philosophical approach will have to focus on who can do the job best and accomplish the most with the available skills, abilities and knowledge and resources. Intuitively management knows that a loyal, well-trained and educated employee is the key to a successful company. But at the same time they must be convinced that the employee is a true partner in the company, supporting and developing the products for the good of the customer. This could be a concern whether the company provides a service or system for their use. This is probably the hardest concept for the management (but not the leader) to accept and translate for all to understand and know. Most managers see their employees as a 'pair of hands' to do the work. They often fail to see the partnership necessary to influence positively the development of the work product.

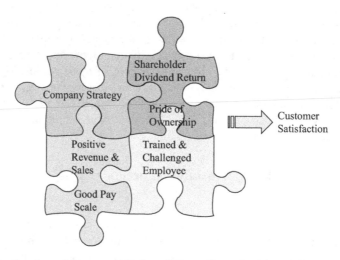

Chart 2.8   *Leader – Managers' Point of View. Note: See Appendix section for full page image*

To establish a positive climate for the development of a truly effective product it is important to have the employee put all their skills and abilities squarely behind the management's ideas, efforts and plans. The author researched this concept for a model that might be used to support the CMM protocol, and the ELITE Leadership Model presented earlier definitively answered that question. The approach that is needed in order to formulate a relevant and true planning process that anyone can use is inherent to these models. We were also fortunate to find a model developed by the Society of Automotive Engineers (SAE) that has been widely used since 1994. The SAE has developed and supported many models and standards for a variety of engineering fields, especially automobile and aeronautical systems worldwide. This model is called the 'SAE Total Quality Management Process Map' and was first introduced through the *SAE Notes* in December of 1994 [6]. Over the years it has continued to grow in relevance and is used by engineers throughout the world. It has become the mainstay in those engineering fields that insist on customer satisfaction.

Starting from a focus on customer satisfaction the process map develops a planning approach around seven fundamental ideas. These are:

1. select the critical few missions/goals/strategic issues,
2. identify and document your key customers,
3. identify and document your customers' key needs based on real data,
4. identify and document performance measures and key processes to support those needs,
5. document opportunities for process improvements,
6. develop and document action plans and
7. implement, monitor and recycle [6].

This model is illustrated in Chart 2.9.

The model provides a viewpoint that supports the customer, employee and shareholder, as well as the ELITE Leadership Model, towards an appropriate strategy for success and planning. It also represents the often-overlooked importance of customer satisfaction with the product and the importance of keeping the customer in the loop to support the product as well return to purchase or secure the new product or service when it is developed.

Each of these items in the process map has been developed to provide the customer with the best return on their investment. When each has been applied, the company also finds it easier to please the customer [6]. Notice as you review this model that there is a leadership component that easily refers back to the ELITE Leadership Model for the organisational, operational and self-leadership issues. Each of the leadership items has a fit with the SAE model and can be seen on an emphasis basis. In addition, the CMM also has a place in this model when the emphasis can be seen to fit the needs of the customer and the output for the product.

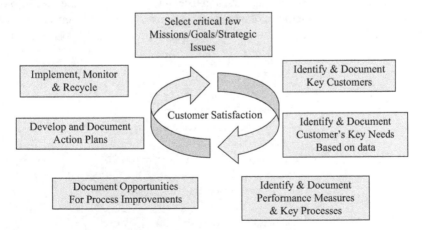

Chart 2.9   TQM Process Map (SAE Notes, December 1994) [Source: Copyright SAE International. Reprinted with permission]. Note: See Appendix section for full page image

2.6   Developing a management plan

Planning has been proven to reduce manufacturing, production or development time by 50 to 75%. This has been proven over and over again in all kinds of engineering and management research at numerous universities. The classic negative situation the author has seen is, 'We have no time to do the planning, but all the time in the world to fix what doesn't work'. We all know this statement is not true. Why is it that we continue to fail to allocate the time to planning or establishing a well-thought-out plan before we execute? Either many of our leaders have failed to get the information about the planning phenomena or they still don't believe it. What we are suggesting and ultimately must do to get the message across to our leaders in management of industry is to allocate planning time to an operation or organisation as it is shared in this chapter. To be successful management must see the benefits of planning and allow leaders to provide a focus on the elements of the process map described above.

The key words in the process map (Chart 2.9) are 'Select the Critical Few'. Missions, goals, vision, strategies and objectives are the critical few from which to choose. Satisfying the customer means understanding their real needs. As most buyers look for products that satisfy those needs, are they durable over time and are they as reliable as they need to be to work every time the product is required to do its thing. Selecting the critical missions, goals and strategic issues will force management to consider only those things that have to be done, as opposed to what one might consider 'the nice to haves'. For example, some of the things we don't often think about are the effects of many of a company's administrative operations on the customer. When we force these activities upon the customer, what type of problems or difficulties are we creating that affect the sale of the product, its use and its resulting distribution to the interested population? Putting a manager with a checklist in charge of the customer service bureau is a sure way to target the company for going out of business. The customer has to be put first and listening (customer service) needs to do just that, listen to the problem or complaint and attempt to come to the customers' aid with some sort of resolution. Saying 'I'm sorry but we can't help you' may be a

> • Based on a vision of the company's future
>   – A blueprint for the action & results
> • Everyone should be involved and informed
>   – Everyone should get involved
> • Based on specific and measurable processes
>   – Real inputs and outputs are discussed
> • Objectives at every level support overall
>   goals
>   – Fulfilment of the company's goals

Chart 2.10   What a Plan Should Be. Note: See Appendix section for full page image

solution for the responder, but helping the customer to find a true resolution is the real solution that keeps the customer coming back for more of the company's products.

The administrative operations or procedures in question might be the company's order entry, credit, billing and order processing systems. One has to ask, are these processes causing problems for the sale of the company's products? How often do we ask similar pertinent questions about the internal processes? Can these processes be improved from a customer viewpoint? How about the location from which we disperse the product? Should other outlets or means be assessed? Has the marketing group done an effective job of assessing the outlets for an appropriate company and product image? Can the product be provided to the client in what they consider a timely fashion? A great example of a good outlet is Amazon.com. They attempt to handle the products their customers are looking for and they make it easy to order and reorder new or used products.

So, the very first step to take in this model is to look at all the mission statements, goals and strategic issues that have been identified. What are the critical few? Focus on those with the right employees to set the climate for success and completion in the planning period. The second step is to identify and document the key customers to whom you are trying to reach. What do we know about them and their needs? Are there things that we are doing that frustrate them, and are there things that they would like us to do that we have ignored? These items should be added to our requirements package and clearly communicated to those who are dealing with the product, the customer and the other ancillary connections to the lifecycle of the product. That's why the third step in the process is to identify, gather and document the customers' key needs. This is important data and it should be shared with the affected stakeholders as well as the personnel and those who can help the customer with this information.

Now that we know what needs to be done and to whom we are going to be doing it, we need to take the fourth step and establish the identity of the performance measures. What do we consider to be meaningful performance issues and how will we measure them? These activities and results should be documented and posted alongside the key processes so that we know what we are doing and how well we are doing it. If we don't know what we should be doing, then that is what is known as 'Not knowing what you don't know'. It is from the meaningful performance measures that we begin to identify the opportunities for improvement. As those opportunities present themselves we should be documenting them and assigning people to take on the roles for meaningful process improvement.

With that in mind, the variations discovered as a result of the processes should be given consideration for improvement; what are the value-added processes, the mainstream processes and the support processes? Of course the value-added processes should be given the highest priority while the activities in the mainstream and support processes are given a cursory look to make sure they are really needed. This can be done using a 'value stream analysis'. Remember that only some of the mainstream processes are providing some value-added activities, and each of these have varying value. The highest value goes to manufacturing and product design (which are the value-added processes); billing, shipping, stocking, receiving, purchasing, order entry and credit evaluation are of lesser value. However, these are mainstream processes and should be scrutinised to reduce waste while improving customer satisfaction. The support processes require even more severe scrutiny; these are accounting, personnel, facilities, finance and the company cafeteria. While systems is often considered a support process, it should be given more latitude than most leaders or management realise: it coordinates the processes as a support, focuses on QA, establishes the mitigation plans for risk and helps reduce activities that are wasteful and cost ROI. Systems to the leader should be a key operative in the organisational structure. It is productive to the organisation and serves as an organisational skill required by the leader and the company.

2.6.1   Adding value to the value-added processes

What can you do to add value to the value-added processes of product design and manufacturing? Adding value to the value-added processes would include a focus on customer-driven design, rapid response to a customer need and the manufacturing involvement in the design where design for manufacturability is paramount. Reduction of waste, reduced set-up times, short time from design to product, and rapid inventory turn-around, order demand and order response would all be good examples of manufacturing issues to be reviewed for added value.

The best way for a leader to instigate the need for review of the value-added processes is through training and education of both management and operations/ design personnel. Each must be informed and educated in why evaluation and review will help manufacturing and design become better, more efficient and more cost effective. The actual activity of reviewing the processes and looking for improvements does not come naturally; it has to be developed through coaching, mentoring and teaching. The participants in this process also need to understand that this is expected by the leader, management and the company with the aim of cost effectiveness. The most common development programme is that of understanding variability in the actual production operation. This development assists the company in providing its philosophy, viewpoint and expectations, which are often only assumed to be understood by the employee, but are not really explained or provided in anything other than a printed handout that is never read. Developing the supervisory staff and production staff to understand variability, how to control it, while instilling the company philosophy is the best expenditure of company time that can be provided. Yes, this activity will cost money and production time; however, in the long run the savings that are realised are double the cost of actually doing the training.

Mainstream processes also need to be assessed. These processes have gone for a long time without evaluation, basically enjoying a life of total abstinence from examination or even simple improvement. These processes are order entry, credit evaluation, purchasing, receiving, stocking, shipping and billing. In many companies they were established as a means of control, to manage and reduce business losses. Today we can no longer allow them to simply operate at will without examination. These processes, often considered simple, must become part of the overall BPR, to reduce waste that detracts from the customer's satisfaction and to improve the efficiency of service to the buyers. Each of these processes, without exception, must become more effective and efficient to provide customer satisfaction. Again, it must be said that these are not naturally assessed processes; it is only through coaching, mentoring and training that the skills and abilities will be developed for the employee to do what is necessary and to bring improvement to the mainstream operations. What one is looking for is variability and detraction from the central theme of meeting the customer's needs. Learning to recognise this in the everyday work structure is not often understood or appreciated by mainstream stakeholders or process workers.

Support processes can only be described as a mixed bag. Some, based on control bureaucracy, were established to serve as control systems for management, while others were put in place to support the company's employees. Where they were established as controlling operations, the skills and abilities are not easily acquired for what are now the necessary efficient and lean budgets with which we operate today. Efficiency and reduction of waste are again learned skills and effective training and education programmes that include coaching, mentoring and teaching must be put into place to reduce the inadequate approaches one often takes to do the job as well as the need for efficiency. These training and development programmes would focus on appropriate, expected and efficient accounting, finance, personnel, facilities and food services. What can these services do that will improve the results of the product and satisfaction of the customer? What are the expectations of the company and how can these organisations be instructed to 'walk the talk?' Good question – and one that is not easily answered. One source would not be sufficient to provide suggestions. However, the most important suggestion would be to encourage reviewers to look for ways to assess the variations that occur in all the processes. The reviewers must also come from the actual cadre of stakeholders or workers, and from customers who are willing to share their frustrations with the inadequate approaches often taken by a company.

To facilitate a meaningful business process re-engineering programme, the company needs to focus on key process design rules that reduce cycle times. These would include re-work reduction, non-value added activities, poor performance feedback, hand-offs and reduction in work complexity, all of which eat costs and waste the company's hard-earned profits. All are variance-loaded operations. These are also bad for business and reduce customer satisfaction, yet we often just ignore the effects and continue on our merry way! Research has shown that cutting cycle time by 50% improves a company's productivity by 20 to 70% [5]. Processing orders, turning over inventory and responding to customer needs more quickly all improve the business, customer satisfaction and the image of the company in its clients' eyes. Research has shown that fast cycle companies react to the market place much faster and institute change in their organisations much quicker [5]. Again these are all items that must be taught as they do not come intuitively to employees or management.

In order to focus on the re-engineering process design elements of cycle time reduction and improvement to fast cycle applications, the author would like to introduce some BPR 'Golden Rules' and 'Commandments' for approaching the process effectively. The Commandments are:

1. Design the organisation around the company's core processes.
2. Design the operations for continuous flow of work.
3. Avoid formalised activities as a matter of company rules.
4. Combine steps, integrating low-value into the direct-value steps.
5. Avoid intra-organisational and shared dependencies/responsibilities.
6. Don't design assembly lines, reduce linear formalised dependencies.
7. Design activities to run in parallel paths to speed up production.
8. Don't mix process types.
9. Design a modular organisation – parts to be re-directed as needed.
10. Co-locate operations for product proximity.
11. Design workgroups to be temporary (changing work packages).
12. Develop multi-skilled workers (increase their scope of capability).
13. Place skilled specialists in the line organisations.
14. Give employees access to *all* the information to do their jobs.
15. Create indirect support groups (no daily control over the processes).
16. Give workers most of the decision making authority [5].

A quick review of the commandments will make it obvious that an immense education and training programme will be necessary for each employee and especially the leadership and management to understand the implications of these ideas. What do these commandments mean? What is the employees' role and that of leadership and management in exercising them? What will the leaders and management be doing to make sure that the rules and commandments are working? What can each employee, leader, manager and worker do to make sure things are going as intended and where do they find help? How does the company expect the mainstream and support processes to be improved? How does this improvement translate into who does what? And what are the expected timelines?

From here we begin to develop the value streams for analysis. For BPR to work we need to know what the existing processes (including the sub-processes) are and analyse them for effective and efficient function. We need to discover where they are undermining the company's performance. To improve, the company must break out of its old patterns; it must discover new efficient and effective ways in which to operate. Both value stream analysis and benchmarking can help in the analysis. These both provide a means of structured evaluation that will generate new ideas and new ways to do things that allow the worker to eliminate the old constraints.

2.7 Process is more than following ISO-9000 or similar management plans

So – you're ISO-9000 certified! That means that you have documented processes in place and that you're using them. That is what the assessors say when they leave your organisation while affixing the stamp of approval to your certificate. But real quality comes from being a mature and efficient organisation, and that comes from having consistent and real internal assessment by stakeholders, owners, operators and employees in the application and improvement of the processes. The maturity of an organisation and its operations results from the ability to say that it has assessed its processes and feels it is doing the best it can in the most efficient manner according to its process owners. Maturity is also the ability to say that one is aware of the appropriate processes, has them in place, is using them with consistent review and is studying the value streams regularly to assure that we are operating efficiently and effectively to deliver our product to our customers.

To truly apply the ideas of business process re-engineering we need to establish some 'Golden Rules'. The golden rules are simple and there are only three:

1. Organise by your products. Effective organisations specialise in their products rather than their functions.
2. Redesign the process flow, workgroup structure and individual duties simultaneously.
3. Minimise the number of groups required to complete a product or service [5].

Just putting these rules out for all to see will not constitute the required operations to be successful. Understanding what they mean will have to be taught to the leaders, all management and the employee, in the language that the company's employees and management understand. Again the importance of training and education steps forward and says, 'We can do this and this is how we will do it!'

Process, as shown in Section 2.5, requires everyone to be involved in making each task effective and efficient one. Requirements must be understood and reviewed. The configuration must be in place, all stakeholders must understand why it is organised in the fashion that it is, and leadership and management must bring everyone on board. With these criteria in place and working efficiently, contracts can be let to subcontractors and managed efficiently with the same expectation of quality as for the prime organisation. But this would not be a reality were it not for the tracking and oversight conducted by the stakeholders, workers and leaders. Again, everyone must understand why all of this is being done and each must have a hand in the actual creation and development of the processes, methods and tools required. The importance of the education, development and training operations for staff and the coaching, mentoring and teaching by the leaders in ensuring that this is the result cannot be undersold.

So here we are, on our way to resolving our problems and using the appropriate processes to get work done. Everyone has been educated or trained to understand why we are doing this and where we are going based on the planning and best case scenarios. As we progress we are aware of the requirements and the ability to change them if necessary and we fully understand the configuration. Our subcontractors are on board and we have hired the best. The stakeholders are the best qualified for the roles they fill and we know that the QA people will provide us with the best feedback for effective operation. Our leaders are mentoring, coaching and teaching their employees and looking for innovative ways to improve. At the same time the plan is being tracked and overseen by the best qualified. We are on our way to a quality product, project or service and we know that it will work because our processes are in control and we have the ability to change them through the CCB if necessary.

Questions for the reader

1. What is the significance of Capability Maturity Model Version 1.1 to the operation of project management and the leadership models that have been discussed?
2. In your company, have you identified the product's operational definition?
3. Have you baselined the requirements for your company's products?
4. Is a configuration management plan being used? How is it defined? Does it help to maintain the baselines as requirements change?
5. Are the work packages under configuration management control?
6. Are standard reports on changes going out to the affected groups?
7. How does the Total Quality Management Model relate to the overall application of CMModel 1.1 and a person's individual capability in the workplace?
8. Are quality review reports being sent to the affected groups?
9. What is your marketing group doing to assure product efficiency and sale of the product to the potential customer?
10. Is a documented procedure used for selection of qualified and reputable subcontractors?
11. Are periodic technical interchanges held with sub-contractors?
12. Are the results and performance of the subcontractor tracked against their agreed-to commitments?
13. Will your customer be willing to pay the price of the development you are putting into place with this project, product or service?
14. How stable are the user requirement that you have determined?
15. Are the activities reviewed with the product, project, service leader on both periodic and event-driven bases?
16. What is the relationship between CMM 1.1 and the ELITE Leadership Model?
17. Business Process Re-engineering has had several phases in the various industries. Give your best interpretation of the BPR application, keeping in mind CMM and the ELITE Leadership Model?
18. Is there objective evidence of engineering QA activities?
19. Are the results of the QA reviews and audits provided to the affected groups and individuals?
20. Does the project/product follow an organisational policy for implementing QA?
21. Are the engineering QA activities reviewed with senior management on a periodic basis?
22. What is the difference between operational leadership and organisational leadership?

23. What are some of the fundamentals of operational leadership that allow a person to flourish in the organisation using their personal leadership characteristics?

24. Can you name some of the organisational leadership fundamentals that allow a personal leadership patron to do well?

25. How would you use some of the 'Golden Rules' of BPR in an organisation and to maintain your credibility?

26. Why are the BPR Commandments so important?

27. What does value stream analysis add to the revamping of an industrial system?

28. Are you as the leader asking the key questions that help guide the project, product or service through the hoops?

29. Can the actual schedule, size and cost of this project be compared favourably to the original estimates?

30. Is corrective action being taken when actual results deviate significantly from the plan?

31. Is there a written policy for the organisation that encourages tracking and oversight activities?

32. Are changes to the engineering tasks and activities agreed to by all the affected parties?

33. Are the performance results, open issues, risks and action items reviewed with senior management on a periodic basis?

34. What has your organisation done to assure that planning is done effectively?

35. Has your company done an efficient job of planning and established an effective tracking and oversight process?

36. Are your company's processes causing problems for the products?

37. How often does the company ask similar pertinent questions about its internal processes?

38. Can these processes be improved from a customer's viewpoint?

39. How about the location from which we disperse the product – should other outlets or means be assessed?

40. Has the marketing group done an effective job of assessing the outlets for an appropriate product image?

41. Can the product be provided to the client in what they consider a timely fashion?

42. What do you consider to be meaningful performance issues in the company and how will you measure them?

43. What will the leaders and management be doing to make sure that the BPR Golden Rules and Commandments are working?

44. What can each stakeholder – the employee, leader, management and worker – do to make sure things are going as intended in a BPR operation, and where can they find help?

45. How does the company expect the mainstream and support processes to be improved through a BPR exercise?

46. How does this improvement from a BPR exercise translate to who does what, and what are the expected timelines?

References

1. Humphrey, W.S. *Characterizing the Software Process: A Maturity Framework.* Software Engineering Institute, Carnegie Mellon University Publications, June 1987, pp. 1–20
2. Humphrey, W.S. 'Characterizing the Software Process: A Maturity Framework'. *IEEE Software.* 1988;5(2): 73–9
3. Zubrow, D. 'Putting "M" in the Model: Measurement in CMMI'. Carnegie Mellon University, 2007
4. Hammer, M., Champy, J. *Reengineering the Corporation: A Manifesto for Business Revolution.* New York: Harper Collins Publishers; 1993, pp. 7–30
5. *Quality Process Magazine.* 'The Golden Rules and Commandments of Business Process Re-Engineering'. Quality Process Magazine. December 1994
6. SAE Notes. 'SAE Total Quality Management Process Map'. *SAE Notes.* December 1994
7. Verzuh, E. *The Fast Forward MBA.* 2nd edn. New York: John Wiley and Sons; 2005
8. QSS Consultancy. *Requirements Management Executive Forum.* Detroit, MI: QSS International; 1998, p. 17

Chapter 3

Leadership is guiding a process-oriented organisation

Figure 3.1   Bombers ready (1934)

Organisations function in unique ways; these functions are necessary to meet their goals and objectives, which hopefully lead to effective and efficient productivity and a quality product. Most organisations have good people at the head of their operations or manufacturing centre who know how to get things done and know who they can count on to get it done. They often can execute the function with the finesse of a finely tuned machine. The organisation depends on those people and often hands them the 'keys' to the company, so to speak, because they are trusted and the company obviously needs them. The organisation places faith in these key people for extended periods of time when productivity is good, but when disharmony sets in or the person in that position is displaced due to sickness, death or quick departure (new job), the continuity that is expected of the efficient workplace is lost. How is process leadership transferred from one individual to another? When a key person leaves it is often not transferred at all. Can the organisation establish an approach that ensures the transfer of leadership over time as is required, or do they stumble, retrain, restart and disassemble when things change due to major changes and upheaval? At best, it is usually the stumbling approach that is used.

Top shelf leaders need to know what it takes to lead an organisation in today's complex environment, don't we? Successful leaders do a good job of estimating, predicting and figuratively guessing about the future. Frequently, they are correct or close enough. But it always seems to be a guessing game. Realistically, 'the environment in most companies is basically unknowable, uncertain, nonlinear, complex and rapidly changing' [1].

To improve a company's chance of survival, leaders must create an organisation that is more adaptable to their constantly changing environment. Creating a process orientation throughout the whole organisation provides the mechanism for the desired adaptability and flexible structure. It goes without saying that leaders must instruct, teach, manage, guide and operate beyond simple speeches and slogans, and exhibit a strong desire to understand and lead their company's process flow. This is what is often referred to as 'walking the talk', and incorporates the fundamentals and requirements of coaching, mentoring and teaching.

Case Study: Labour strife at Boeing erodes followership

At the end of World War II, Boeing, in common with all aircraft companies, was facing a tough transition from war production to providing a product that airlines could purchase. There wasn't a lot of money, but there was tremendous potential in the American economy. Everyone was tired of war and wanted to get back into the business of making money.

Boeing's primary product was the B-29 for the United States Air Force (USAF). In late 1945, William M. Allen, the newly appointed President of Boeing, faced a tough dilemma. The firm's factory was full of aircraft that the USAF no longer wanted. Boeing's competition, Lockheed and Douglas, already had commercial aircraft in production, the Lockheed Constellation and the Douglas DC-6. Boeing had nothing and the airlines were lining up behind their competitors.

'[Allen] knew intuitively that the potential was there in the bowels of the company – the experience that his engineers had accumulated during the war. He decided he must hold the force together' [44]. He had to do something, so he started a new aircraft. His engineers went to work and converted the military C-97 cargo carrier, affectionately known in the military as 'old shaky', and created a product, the Stratocruiser, a luxurious four-engine, double-decked aircraft.

However, Allen still had too many employees and as yet not a single customer for the new aircraft. 'It soon became clear that there would be insufficient work to keep the total engineering force busy for long. In the spring of 1947, the Stratocruiser design effort passed its peak. More than 300 engineers – about 16 percent of the force – hit the streets [44].

The Boeing sales force was able to sell all 56 Stratocruisers produced, but at a loss on each aircraft. The company only made money on the sale of spare parts. However, this was only the beginning of Allen's troubles. While Lockheed's operation was a lean, mean production machine, Boeing did not have that sense of feeling in its ranks.

The labour climate began to change in the fall of 1945 when workers were laid off as fast as the company could process their notices. It was simply goodbye and good luck. In September, the union rescinded the no strike pledge it had made at the beginning of the war [44].

Allen's decision was to structure a new agreement with the union. Had he been in touch with the employees, like Gross at Lockheed, he might have been able to read the mood of the crew better. Allen asked the union to open negotiations for a new working agreement. In a letter to all the shop employees, he wrote, 'the present labour relations agreement has become unworkable to such a degree as to seriously impede progress of the company toward peacetime production and maximum acceleration of employment' [44]. Not exactly a cooperative climate in which to start negotiations.

Many supervisors, who had moved out of the union during the war as production built up, wanted to move back into their labour positions. The union resisted. The courts sided with the unions, and Boeing was forced to lay off 670 supervisors. 'On November 15, the remaining supervisors did not report to work, and production nearly came to a standstill. Allen took direct action, sending a personal letter to each of the striking supervisors. Most of them returned to work five days later' [44].

'Later in the year, Boeing Engineers in Seattle, Washington (USA), formed a collective bargaining organisation (SPEEA). They declined to call it a labour union, and signed an agreement with the company. In a National Labour Relations Board (NLRB) election, SPEEA was certified as the bargaining agent for the engineers' [44]. Lockheed engineers later formed a union at their Burbank location.

By this time the culture was beginning to be set. The union distrusted the company and vice versa. The engineers didn't trust either party. When it was time to renegotiate the union contract, the company wanted more flexibility in movement of its employees. They basically wanted to be able to move people from non-union positions into positions previously held by union members. 'After the concessions were formalized, the proposal was put to a vote at a mass meeting on 24 May, 1947. It was rejected by a 93 percent margin. Immediately after the vote there were cries of strike (by the union members)' [44].

This was not a good time for a strike. Boeing was still trying to establish itself in the commercial aircraft market. There were plenty of people for jobs, so the union was not in a strong position either. In contrast, 'In California, workers at Lockheed agreed to a new contract for less money than Boeing workers were already receiving, and the SPEEA engineers signed their new contract' [44].

It wasn't a sound economic decision to strike, but the union did not trust the management. The seeds of strife had been sown several years before. Tensions were high. This was a bitter fight carried out by the union between April 1947 and October 1949. It created dissent between management and the union that could not be resolved. It soured the culture. In 1950 the company and the union signed an agreement. Part of the agreement was a clause that stated that new employees could join the union or not, as they chose, but having joined they must remain members during the term of the contract. This loss of a 'union' shop hurt the union's bargaining power.

Boeing beat the union but lost its family and its potential followership in the process. 'On 22 May, 1950 – more than two years after the workers had gone on strike – a one year contract was signed with lodge 751, ending the longest and most bitter confrontation in Boeing's history. Seven months later, in Wichita, Kansas (USA), a similar contract was signed by Lodge 70. The strife was behind them but the scars have remained' [44].

Questions about this case:

1. In your personal opinion, what was the reason for the professional engineers union that formed at Boeing?
2. Boeing made a different decision from Lockheed in its involvement with its employees and the unions. What do you think Boeing could have done to improve its results and why did it differ from the approach used at Lockheed?
3. As a process for research, study the current conditions at Boeing and Lockheed and provide your team with your assessment of the conditions and why you think these factors are valid.
4. In today's aerospace industry, there has been a total turnaround of who is in the commercial aircraft business versus those who are currently in the military aircraft business. Why do you believe this has taken place and what may have led each company to go in the direction that they have chosen?

To support this change in the current organisational structure, a new orientation is needed towards the actual culture which is non-judgmental and non-blaming [2].

Management must create and reinforce this non-blaming atmosphere. Only if fear of retribution can be eliminated will the stakeholders, leaders and employees be willing to examine their work, with a view towards removing waste and driving down cost to develop a quality product. If employees are protecting themselves and their livelihoods from an intimidating management environment, they will never scrub the operations hard enough or at all to remove waste. Retribution and intimidation threatens their safety net. Management must demonstrate to leaders and employees that a safety net is provided in the organisation and that without question they can safely examine costs or any other waste-producing factors. Only if employees trust and truly believe in their managers, leaders and supervisors will they feel safe to scrub cost and drive out waste. The more people in the organisation who share this belief, the more effective and profitable the organisation will be. If this attitude is promoted on the shop floor or production/engineering areas, but ignored or even laughed at in the offices of finance, human resources, marketing, etc., the message will get out that the leadership is playing a game and is not really serious. 'The structure needs to be compatible and supportive with the culture and both should be ecologically matched to the environment and the organisation's purpose, strategy and vision' [3].

Leadership and management must educate those in the organisation in the value of looking at the company as a complete and whole system, with all its components, parts, departments and divisions interconnected and interdependent. This can best be done by encouraging coaching, mentoring and teaching by the leaders and managers. Organisational systems flow through many types of processes. It does not matter what the company produces or what services it provides. All organisations are composed of productive processes. And these processes must be seen as interconnected and productive or the organisation will die and cease to exist.

It takes a leader with an operational process orientation and a people-oriented supportive focus to move a company away from a bottom-line, financial focus and towards an organisation that provides policy support for operational processes. 'Leaders that lead by collaboration, processes, compassion, communication and values, not by planning, organizing, directing, staffing and controlling; these are the leaders capable of bringing energy and understanding to the local challenges and at the same time integrating those local actions with the organisation's purpose and direction' [4].

Most companies are driven by the desire for results and profit through production of a product or delivery of services. This was no different for the management team responsible for the *Columbia* flight. Rightfully so, as productivity is essential for the survival of the product and the company or organisation it supports. However, many companies, once they set up a department and its processes, ignore the operational and organisational leadership orientation required and proceed to cajole, harass and badger stakeholders or employees to increase their productivity through whatever means they have available. The appropriate process used by the employee to produce the product, as well as the correct tools to be used, is often ignored. Only results, the development of the finished product, seem to count. Most employees get their work done, ignoring official processes and procedures, and bow to the needs of quotas set by the department silos, organisations and others in power. 'The informal networks, the practical decisions and actions, and the common sense in doing a job end up driving the day-to-day operations in most organizations' [5]. 'Management needs to shift its focus to improving the process employees' use and make it easier for them to be productive' [6].

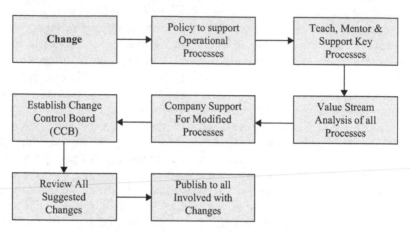

Chart 3.1   Change, Policy & The CCB. Note: See Appendix section for full page image

Again, leaders need to create:

> an atmosphere of trust, open communication, collaboration and freedom from fear and reproach. While there are managers and supervisors, leaders need to be less autocratic or controlling and behave more like mentors, teachers, coaches, colleagues and supporters. [Leaders need] to take on responsibility for projects, products or services and at the same time give colleagues the freedom to think creatively and have a strong voice in their own tasks or activities. Leadership in the action culture [today] is very different than the classical models offered in the past. [7]

To survive in today's turbulent, ever changing environment, an organisation must, through knowledgeable leadership, be more flexible, adaptable, intelligent and accepting of the complex requirements than in the past. Changes in the environment drive this new requirement. 'Complex adaptive systems cannot be controlled, they can only be nurtured. Control stifles creativity, minimises interactions, and only works under what is considered to be stable situations. It is not possible to control a worker's thinking, feeling, creativity or trust' [8]. Today there is more information available than an individual can comprehend and definitely more change than any leader or manager can adjust to or control. Organisations that don't adapt using new approaches will not be able to operate in the new and emerging environments of the future. 'Only knowledge can provide the understanding needed to deal with this complexity. Such a milieu demands a different paradigm and the use of new rules and roles for leaders and managers' [9]. This is why the author believes that the ELITE Leadership Model provides the best answer to providing the gateway to the tools required to be an understanding and successful leader. The successful leader of the future will have to develop the skills required of the model if they are to prosper.

3.1   The fundamental flaw of 'heroes'

The fundamental flaw of the 'hero syndrome' is that many in the organisation rely on the person in the 'key' position and not the documented or proven process established to accomplish or complete the required tasks or activities. When the 'key' person vacates the 'hero' position there is nothing there to carry out the 'hero changed' processes that they put into place over time, especially once they are gone. In addition, as the 'hero' changes the process over time and we find they didn't document those improvements or changes, the true, currently executed process will be lost once the person leaves. The company is then left with only the know-how to do things the old way before the hero changed the process. A company must, if it is to succeed in the long run, develop and support a policy that states that changes to processes must be documented; or even if said process is to be used at all. The demands on stakeholders and employees must resonate with the requirement that if you can't find the process you are required to follow, then you shouldn't be following it. If that new instructed process makes sense then it should be documented, established and followed based on a company policy and on Change Control Board (CCB) approval. The company leadership has the responsibility to insist on this action and must teach, coach and instruct employees on its importance both to the employee and to the operations within the company.

Leadership in this new dynamic environment requires the old command and control styles be abandoned. Leaders must balance the needs of the team for autonomy and flexibility and the needs by the organisation for control of the resources. In a very dynamic work environment decisions need to be made as close to the customer or company work environment interface as possible. The primary contact is the stakeholder or employee dealing directly with the customer to improve the product, the customer's impression of the company and the product's ability to do what is advertised. Ideally, critical decisions should be made as close to the customer as possible so as to maximise customer satisfaction. However, the leader needs to manage the resources used by many specific decision makers, which is most appropriately the use of changes needed to the processes to improve the product. This must be done in an acceptable manner to the company and a recognised approach that has been established as policy and acceptable to all members of the stakeholder community. All members (managers, supervisors, leaders and employees) want to maximise their self-actualisation while fully moving towards the organisations' strategic goals, again an appropriately higher productivity level and a reduced cost for the development of the product. Those opposed to this are those who, deep down, don't trust others to work in what they would consider to be the correct or appropriate process and direction for the company that gives individuals real responsibility to the role and the activity.

Organisations function in many ways to achieve a quality product or service. Most organisations have operations people who know how to get things done, know who they can count on to get it done, and are able to execute the function on schedule and at cost. The company depends on these people because they are trusted and, more importantly, needed. They are relied on to keep things rolling when times are good. When disharmony or dissonant change sets in, they often lose the continuity required to maintain productivity. 'The challenge of the new leadership is to create, maintain, and nurture their organisation so that it creates and makes the best use of knowledge to achieve sustainable competitive advantage .... Collaboration is such an important part of creating the right environment and leveraging knowledge' [9].

Can we establish an approach that ensures the transfer of leadership over time as is required, or do we fumble, retrain, restart and disassemble/reassemble when things change due to a major upheaval? Knowledge about a company's operations can no longer be transferred only in the traditional ways or methods once used, through books, in a classroom or unrehearsed on-the-job training. Instead, it must come through on-the-job cross-training of team members that will ensure that individuals know and are able to do each other's jobs or activities and act interchangeably. This is why it is important for a leader to understand their role as a coach, mentor or teacher. The leader has to 'walk the talk' for real and continue to do so for as long as they work in that environment. By example the leader shows the employee the way and coaches them through the correct path for the event. The transition of work, its activities and knowledge, need to be dynamic, planned and executed in a timely fashion by leaders or managers.

New leadership tactics must encourage and require a culture to be in synch, and one in which process changes are documented. The importance of establishing the policy and procedures for appropriate process change and the use of the CCB are therefore unquestioned. Leadership must emphasise to employess that if they cannot find current documentation of the process they are required to follow, then they should follow the existing documentation or not do it at all until there is a clear delineation and acceptance by the CCB and leadership of the new process. The employee should then follow through encouraging the change and its appropriate acceptance so that it is documented and accepted by the company. If the current proposed or new process being emphasised makes sense then it should be documented, established and followed. The CCB can make it so, but only if the change and new process is brought to their attention.

A traditional command and control structure exists for organisations that are stable, predictable environments. Unfortunately, that environment exists less and less in today's businesses. Even the support areas of most organisations that do not directly interact with customers must be flexible and ready to adapt to changes driven by those who do interact with customers and the competition. Again, when changes are discovered and need to be implemented they must be documented and available for all concerned so that action can take place.

'Once upon a time, heroic leaders steered an organisation with a firm grip and solved problems single-handedly while still managing to keep the troops inspired. For better or worse, that stereotype doesn't fly anymore in American business' [10]. Traditional companies were once hero-dependent. The successful leader who has reaped the rewards of the past must now change if they are to be successful in the current business climate and most definitely for those who will operate in the future. Without this new philosophy, a company will be mired in the past and suffer loss of business loss and even bankruptcy.

The classical autocratic leader cannot be successful today, because no one individual knows enough about the potential of business to second-guess the changing environment. The charismatic leader would not be long successful, since knowledge workers, while inspired by passion, are rarely taken in by surface glitter and personality. Strong, individualistic leaders want and expect control and visibility, since they lead by personality and image. [Leadership is needed] throughout its structure to aid in cohesion and rapid adaptability. In addition, the [organisation under these conditions] cannot be designed and constructed; it must be nurtured and allowed to co-evolve with its environment through self-organisation at the local level and iterative interactions with the outside world. This is not something strong ego driven leaders are good at, or willing to do. [11]

New leadership is required for future growth. A new type of leader must have integrity. If their staff do not trust them, then they will not follow and the leader will fail. This process is known as positive followership. These are but one of the personal characteristics of a leader. The leaders must also be able to visualise the future through the company's stated and published goals, objectives and communicate their vision of them for the understanding and future of the employees. Leaders must support their employees, encourage them, reward their successes quickly and help them improve to overcome their deficiencies in a timely fashion [2]. If we go back to the ELITE Leadership Model (Chart 3.2) we find some of the answers to the questions that plague us in search of leadership [40]. Knowledge of the self is primary. It is almost like a foundation where our self-awareness, self-management, social awareness and social management serve us well if managed appropriately. But these are skills that have to be developed. People leadership builds on that foundation using our ability to develop and build effective teams while understanding our responsibility to develop people, motivate them and demonstrate the managerial courage to lead in the face of adversity.

Much of this we should know. However, the next two components are not so well-known and are only learned in most cases through the 'school of hard knocks'. These are operational leadership skills and organisational leadership skills. Operational leadership focuses on process and the appropriate management of the key or core factors. It embraces the rules of project management and systems thinking. While the candidate is aware of and knowledgeable of the business and the acumen that enables a company to operate with seasoned judgment through experience, they must focus on effective process management. Organisational leadership understands the importance of an appropriate vision, strategy and mission for the company. The organisational leader understands the need for change management from the enterprise perspective and works hard to keep the organisation focused on the customer and a quality product.

Chart 3.2   The ELITE Leadership Model [Source: Reprinted with permission of the University of Tulsa, ELITE Program]. Note: See Appendix section for full page image

Tomorrow's leaders will have different skills and values to today's. They will be 'admitting that they may not know more than their knowledge workers about any given problem, and trusting in their people to think, (along with the ability) to do the right things' [12]. Leaders of the future must create a new work atmosphere of trust, integrity and confidence in their employees. Future managers and leaders must also create an environment where mistakes are understood, learned from and tolerated. They must realise that 'freedom to make mistakes is the price for creativity, agility, learning and optimum complexity' [12]. They must support and encourage their employees while still being held accountable for the organisation's results and their employee's success.

As the business climate requires faster and faster change, organisations must adapt to it or go out of business. The 'John Wayne' hero who resists change forces the organisation to fall behind in its market competition. Not that John Wayne wasn't a 'hero' in his day, where the environment meant stability, where command and control was king. No matter how large a company may be today, it can never afford to ignore the market or its competitors. It must adapt accordingly. Continuing to operate with an antiquated approach to management and leadership will only put an organisation that much further behind the power curve.

For most of modern business history, managers have worked to prepare companies for the future by providing guidance and advice to their employees on how to operate in new environments. They have tried to anticipate customer demand and

- Self Awareness
  Development
  - Self Awareness Skills
    - Self Awareness Skills
      of Effective People
  - Management Skills
  - Social Awareness
    Skills
  - Good Relationship
    Development Skills

- People Leadership
  Development
  - Effective Team
    Building
  - People Development
    - Coaching
    - Mentoring
    - Teaching
  - Motivation Skills
  - Functional Courage

Chart 3.3   Self Awareness and People Leadership. Note: See Appendix section for
full page image

defend against competitors and the resulting changes. Traditional management philosophy is that senior management always knows what is best for the company. Management would then provide the strategic guidance their employees needed to make the company successful:

> The often unstated assumption is that the future will be like the present, or at least that the future is predictable and they have a good idea of what it will be. Senior managers would consistently look for another new idea, work a little harder, and continue to do what they have always done: identify the gap, write the strategy, and implement according to plan. [13]

Case Study: Loss of key persons

The programme is completed. The company is closing down the accounts. With no more funding in the accounts to charge the engineers' time to, it is the company's opinion that it is time to lay off the personnel. Managers work to protect the few good people they can afford to keep. But most unnecessary employees will either be laid off or transferred to other programmes. This process of controlling the costs in aerospace programmes has worked as long as have been new programmes in sight or over the horizon. Survivors in the aerospace industry have very interesting resumés that list a variety of unique and different programmes.

The classic situation in the industry was the end of the space race to the moon. When NASA cancelled the last moon shots and drastically eliminated staff at Cape Kennedy, the town of Cape Canaveral, Florida went into a depression. Engineers in that location could not find jobs. Some were forced to work at fast food restaurants or become taxi drivers where jobs were open. Many finally gave up and moved out of the area. Many could not sell their homes so they just walked away and left the keys in the front doors with a note inviting anyone who could take over the mortgage payments to move in. It was a sad time, and a really bad time for the aerospace field.

Today, things are a little different. While there are still fewer new aerospace programmes starting up, other technology fields, notably information technology, are attracting qualified engineers away from the feast or famine world of aerospace. As a result, when new aerospace programmes do come along, qualified engineers and professionals may not be available. An example of this phenomenon occurred in Colorado, USA. A major aerospace company had just completed one project and significantly reduced its headcount. When a new contract was brought to its attention, the company was not able to hire sufficient engineers to fill the roles. As a result, it overworked their onboard employees and still missed set deadlines. The boom or bust cycles in the aerospace industry had relied on a readily available supply of engineers, and this was no longer the case.

Two employees, Jim and Matt, were both caught up in this rollercoaster ride. Both were aerospace engineers with 10 to 15 years' experience, respectively. Jim had been going to school at night, working on his Masters Degree in software engineering. He knew aerospace was always unpredictable and the software business was booming. Matt, however, had been with the company since graduating. He liked what he did and had no interest in any field outside of aerospace. They both received lay-off notices as their programmes ended. Matt was now in a quandary. He didn't want to leave the field; however, the company was the only aerospace industry in the town. He was forced to look elsewhere. Jim decided this was a good time to jump fields, so even though he was without the degree his experience allowed him to interview with others. He found another job as a systems engineer with a software company in the same town. Matt was unemployed, relying on his wife's income. Ultimately, had to move to Kansas to relocate with another aerospace company [43].

Questions about this case:

1. Do you think the company could have done more to retain its valuable employees?
2. How do you think the boom and bust reputation of aerospace affects companies' ability to attract new professionals?
3. Developing good professional engineers and scientists in the field of aerospace is a complicated effort. Can the industry continue to afford to lose this valuable experience?
4. As employees of the particular industry in which you are employed, can you continue to concentrate on a single professional expertise? How might your plans change considering this case study?
5. Must we now be prepared to jump from industry to industry depending on a company's programme lifecycles?

It's not that management was wrong, it's that the environment outside the organisation has changed and operates differently from the past. Organisations can no longer be operated from the high atmospheric levels of senior management. Their reaction time is inadequate to respond to environmental changes and their profit and loss statements are inadequate and often too slow to provide the necessary guidance needed for the company to survive. The complexity of business and the production of goods and services far outpace the knowledge any one manager can develop. Teams today are made up of stakeholders, managers and leaders, and often include all the necessities in the fast-moving climate that can help it succeed in business.

'The world has become both increasingly complex and increasingly transparent. To be credible in the business world ... leaders need to respond effectively to that complexity, while also being more transparent about the reasons for their decisions and communicating with an extremely diverse workforce' [10]. Due to the speed of change, and the current need for rapid communication, leadership and management can no longer make decisions in smoky backroom meetings based on their past experience. Leaders need to use all their resources to make the best decisions. These resources include the knowledge workers on the team, and the wisdom with which they surround themselves through their colleagues and fellow workers. Including stakeholders and employees in the decision process requires many approaches to data analysis and evaluation. The leader must trust the people surrounding them, and especially the stakeholders, using good reasons and analysis. They cannot afford to leave out any talent in the evaluation, review or analysis phase of planning. Collaborative leadership using all the resources available must be seen to replace the command and control approach as the new method required to process the best decisions. 'These (types of) leaders are seen as equals, but equals who help and assist others to get their work done, equals who are good role models to mimic, equals who are about others personally ... knowledge workers need to do their jobs' [14]. The stakeholder probably knows their customer's needs, values and wishes, which are far better at predicting strategy, direction or approach than the existing and antiquated management, who do not have daily contact with the customer. Gone is the 'dark brooding, stern-looking captain' standing at the helm of corporate command. Communications between leaders and followers cannot be slowed, but must be faster and more transparent. Leaders are invincible only until they are shown to be human.

Future leaders must focus on helping – by coaching, mentoring and teaching – not controlling their employees. Because of the dynamics of change, and the knowledge requirements of today's environment, employees are now, more than ever, in control of the organisation's productivity. Leaders have become facilitators:

Their objective is to leverage their onboard competency, the employee capability and not direct it. They are participants in the process, [not the all seeing all knowing] directors. They combine the art of collaboration and the art of leading others. [15]

The new roles for leaders are to assist, encourage and facilitate their employees' efforts and provide them with whatever resources they need, to maintain a position on a learning environment, and to keep everyone focused on the organisation's strategic objectives through coaching, mentoring, teaching and assisting.

The power games and efforts of top management to control the organisation have been replaced by a supportive, evolving team culture. Employees 'with their creativity, initiative, loyalty and competency – without exception represent the single most valuable resource' in the company' [15]. The leaders' role is to support them and provide whatever resources they need to maximise their productivity. This does not mean managers and leaders have lost all control and the boat is without a rudder. Together the manager, leader and employee teams now know strategically and daily what has to be done to be a successful producer. The stakeholders are closer to the customer and the competitors and are able to respond to changes faster if the company policy, processes and methods allow them to.

Employees were always closer to the customer, but change was slow enough that delays in the decision process as they went up and down the organisational chain of command could be tolerated. Today, with rapidly changing environments, a delayed decision-making policy, process or method will not allow a company to keep up with its competition or even meet the needs of the customer.

The shift in control must be available to the team responsible for a specific segment of the project, production or service process. The team must have a clear idea of their goals, what has to be done, what resources are needed and the schedule that must be met. The challenge of management and leadership is to allow teams enough knowledge and control to achieve their potential and meet the requirements set for the product. The challenge is in convincing managers and leaders to release their knowledge and control to the teams.

> When push comes to shove most managers will choose control. In fact, it is emotionally difficult, in most companies, even to relax the emphasis on control. Managers and leaders who are doers, accustomed to getting things done, will tend to trust themselves more than anybody else. [16]

This form of management resistance will hurt the organisation's productivity in the long run, maybe even the short run, and stunt the company's ability to meet its customers' needs.

The natural reaction of most managers during times of stress is to increase control. This is typical in what is known as fire-fighting crisis management. Solve today's crisis and worry about tomorrow when it arrives. Most managers and some leaders, especially those at the lower levels of the organisation, do this on an everyday basis. The problem is that the outside world is not going to wait. If anything, a business is going to pass those that have slowed or stopped for adjustment and will move ahead through a better approach.

- Problems begin & continue because no-one
  teaches the employee how to control or master
  the processes
- If the employee looks to the leader to solve their
  problems they will not understand:
  - Variability
  - Supplier & Customer Links
  - Process Improvement
- Teach, Coach & Mentor on a real-time basis
  - How are they doing NOW, instead of once a year

*Chart 3.4   Mentor, Teach & Coach. Note: See Appendix section for full page
image*

For the leader to have credibility, it requires the capability to self-organise, adapt, and respond rapidly to changing events; knowledge workers (and their teams) are empowered to use their knowledge and act, sometimes on their own, more often within teams. ... When push comes to shove, most managers will choose control. [16]

This approach will have a devastating effect on today's companies as most employees see themselves as being more than able to solve problems or to offer answers.

If an organisation is going to operate in a more open fashion, as it must, it has only to recognise that several cultures exist in companies today that must be understood and appreciated. Let me just name a few of the cultures. As I do so the self-recognition will astound the reader. Each of you have heard of them and know of their existence, but have you thought that they matter in your organisation?

Most of us reading this book are familiar with the first generation, the pre-boomers, the veterans, the silent generation, the traditionalists, the seniors. Yes, they are all the same group – those born between 1922 and 1945. They are in many cases those that took us under their wings and taught us what we currently know about leadership, efficiency and management. The world events that they experienced shaped their world and the mannerisms that they portrayed, especially in leadership and management. They were the 'John Waynes', the 'Audie Murphys' who knew a lot about command and control, because that was the way it was done in that day and age. The next generation was the baby boomers, the me generation and the sandwich generation (born 1945–60). Again, all the same time period but a totally different culture and set of attitudes. The next is generation X, the cuspers, and the buster's (born 1960–80). And last of all for this discussion are the millennials, generation Y, the nesters, and the echoes (born 1980–2000). All of these differing attitudes, cultures and ideas within an organisation make up a melting pot of approaches, ideas and methods for getting things done. We know that command

and control will not work anymore because of the rapid pace of business and the integration of their ideas and beliefs. Therefore the team approach must be taken so as to account for the various attitudes and cultures present. The true leader has to understand the talents and belief structures of whom they are working with. [17]

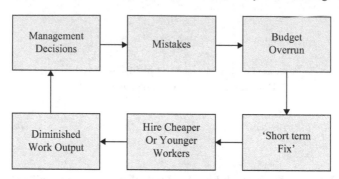

Chart 3.5   Experience-Based Negative Feedback Loop [Source: From Frappaolo, C. 'Consultants View: Building a knowledge management program'. Beyond Computing, 14 September 2000]. Note: See Appendix section for full page image

'Toyota created a culture that enabled every employee to participate actively in their operational improvements every day. That is the challenge that faces all companies today' [2]. The company 'monitor[s] the morale, frustrations,

Characteristics Also known as: Born Between:	Pre-Boomers, Veterans, Silent Generation, Seniors (1922–1943)	Boomers, Me Generation, Sandwich Gen (1943–1960)	Generation X, Busters, Cuspers (1960–1980)	Millenials, Echos, Nexters, Generation Y (1980–2000)
Work Habits:	- Follow Tradition - Status Quo - Obedience over Individualism - Advancement Through Hierarchy - Sense of Duty & Honour - Natural Leaders	- Value of Personal Growth - Wants to be Involved - Team Orientation - Value Company Commitment & Loyalty - Sacrifice for Success - Uncomfortable with Conflict	- Entrepreneurial - Independent - Thrives on Diversity - Desires High Level of Responsibility - Constantly Looking for Creative outlets - Quickly Moves on if Employer Fails to meet Needs - Impatient	- 24/Seven - Capacity for Multitasking - Global Connections - Competitive - Civic Minded - Diverse - Desire for Structure - Goal & Achievement Orientation

Chart 3.6   Culture Differences [Source: From Kennedy, M.M. 'Career Strategies'. Presented at ASEE College Industry Education Conference, ASEE, 2007. Also available at www.moatskennedy.com]. Note: See Appendix section for full page image

perceptions and attitudes of their colleagues and makes adjustments accordingly. The organisation's sensitivity to informal networks and to cultural changes, coupled with their collaborative style and leadership perspective helps (the manager and employee) to nurture the culture ... ' [18]. Toyota faced the same problems we face today with a changing mix of generational cultures. The secret was to establish a corporate culture that recognised the needs of each generation and allowed them to practise and excel in their own way while making a valuable contribution to the organisation and its product. The company instituted teams that allowed for input and in many cases control of the assembly line itself. It established a system that controlled that change for all to benefit. As astakeholders in the Toyota enterprise employees had the ability to stop the assembly line when they saw an operation that was out of synch. They were also given the ability to have input into the processes and make changes on a daily basis if necessary and even encouraged the posting of these changes.

3.2 The evolution from hero-based to team-based

Today's leaders must work with teams. Today, teams 'own' their processes for production. To get the maximum out of each team, leaders must create a certain amount of discomfort within the team. Not the degree of discomfort that causes teams to take their eye off their objective and to focus on internal problems. However, leaders should want to encourage teams to continuously improve. They should challenge the existing methods, tools, processes and procedures. If they find a better way to do something, then the leaders must trust their teams to make the improvements necessary. Leaders help teams make these improvements. They provide needed resources of budget, time and schedule adjustment. Once proven the new processes are documented and in follow-up improvements are encouraged even for these changes. It is only through coaching, teaching and encouraging teams to maximize their efforts can leaders maximise the team's performance in today's environment. [2]

Traditional leaders are viewed as: 'calm, decisive and demanding'. They never show uncertainty or a lack of resolve. 'They are pragmatists who favour results over values and believe that the shortest distance between two points is always a straight line. They exhibit personal power in the form of charisma, inhabit power by position and use power as a blunt instrument to achieve their goals. They never admit to failure, and by implication, they never grow. Leaders are prone to keep doing what has made them successful in the past. ... Unfortunately, the past and some of these experiences can be a handicap, if not a prison (for some). The inclination to reach for tried-and-true approaches means that you're effectively blind to the opportunities and hazards of constant change' [10]. There are so many options, interests and cost/benefit comparisons that a single right answer may be almost impossible to find. This is the reality that today's leaders must learn from the cultures of their stakeholders and must practice in the workplace.

This management approach worked in the past, when tradition had value and change was slow. But today's pressure to change to the new forms of competition and the environmental effects of constant change is so great that organisations can no longer tolerate traditional leaders and their habits of past experiences. Today, the changing environments, markets and competition dictate a faster rate of change, a different outlook at how things get done and a better relationship with the leader, the stakeholders and the employee. Without this new outlook a company is going to struggle, fail and find its products are being beaten in the marketplace by many competitors who understand the lay of the land.

If you keep doing what you are good at while the world around you changes, you will eventually become less competitive. You may be very good at what you do, but out of step with the changing markets. A classic example is Sears and Roebuck, at one time a monarch among businesses. Who would have believed that such a powerhouse of a department store would one day be bought out by a Five and Dime novelty store, S.S. Kresge, the change brought about K-Mart, and an entirely new corporation operated by the K-Mart management called 'The Sears Holding Company'? However, at the time they were acquired, Sears had become complacent with its business practices and operations while the world around it changed. Sears and Roebuck had become comfortable with its business processes and profits and thought the world would remain the same. Sears became a name-only business through the ingenious management by K-Mart and surrendered its department stores to K-Mart. The big story is that in recent years, Sears has become important and through K-Mart has reinvented itself. The K-Mart stores in some states are now being operated as 'new' Sears outlets and we are seeing a resurgence of the Sears business in a different and more welcome form. At the same time K-Mart has spread its wings into other countries, such as Australia and New Zealand. In 1970 K-Mart sold its interest in the two countries to the Australian Business entity known as K-Mart Australia for a profit.

Many other organisations have also relied on heroes to push the work through the organisations' bureaucracies (or other internal ownership empires). The hero knew who to call, what would please which manager and how to prepare acceptable proposals and results for key individuals. That personal 'hero method' has become less and less effective in today's business process. Today the company processes are more complex and rapidly changing, outpacing the old know-it-all hero, their ability to stay abreast of the operations and knowledge of the changing landscape. To survive, organisations must now depend on fully functioning teams or process-oriented systems in which they can adapt to the changes in the requirements and the customer need arenas. Effective leaders need to trust their team leaders to carry out the performance responsibility expected of the client. Leaders must work with the teams rather than command them, accept their inputs and adjust to the customers' needs. In today's business environment the team really knows the 'in's and out's' of the productive process, what methods to use, the tools that make it more efficient and the needs of the customers. 'Leaders, change agents, and many knowledge workers must understand their own organisation so they can continuously fine-tune them to match the dynamics of the [business] environment' [4].

For a leader to come into an organisation and attempt to give commands based on past experience, would interfere with the team's productivity. Instead, they must listen, support and provide the team with the resources necessary to move them towards the company's strategic objectives. Another way to think of it is: the team drives the company's engine; the leader reads the strategic map, provides the necessary fuel and repairs and steers the company in the appropriate direction based on this input.

The conditions required for a successful teamflow experience according to Mihaly Csikszentmihalyi's research (as cited in [20] and [1]) are:

- Tasks must have a good chance of being completed, yet not be too easy.
- The team must be able to concentrate on what it is doing. Interruptions, distractions or poor facilities prevent concentration.
- The task should have clear goals, so that the team knows when it has succeeded.
- Immediate feedback should be provided to the team so that it can react and adjust its actions.

As explained earlier, companies must transition to a new method of leadership in order to survive. This transition will not be accepted or occur easily or be welcomed by everyone, especially management. It requires a great deal of understanding, explanation and added work from leaders and managers, key process holders and stakeholders. Each team member must focus their time and knowledge on improving the organisation processes with the help of all. Those who do not understand or appreciate the methods will resist the change and push to maintain the status quo. If the management or leadership only gives lip service to the required methods, it will not withstand the push back from inside people who have a vested interest in the status quo. Traditional managers and controlling leaders will see these changes as threatening and the future plans as full of unknowns, concocting a mess and creating unnecessary adjustments to a system. This is of course all in their minds, because everything as it is works well. They will naturally resist.

Future effective cultures in organisations will have individual workers take the appropriate action at the right place and right time. To be effective, future companies must be 'built on the foundation of creating, leveraging and applying knowledge anywhere, anytime it's needed' [21]. Within these companies, operating in a stable business environment, an effective culture can be built into the best structure and managed well. However, we must see that 'as the environment gets more dynamic, nonlinear, complex and unknowable' [21] controlling this through a leader, management or a team leader will be harder and harder to do'. An example of this condition is the current mobile phone industry, in which new products and service announcements regularly reverberate throughout the whole industry. A proactive company will provide an environment that allows its employees to learn what is necessary when they need it and attempt to keep up with competitive changes. This condition cannot be mandated from top management, but must be encouraged and desired by stakeholders, leaders and employees with the appropriate leadership from an understanding member or leader of the team.

Management must work to provide whatever information the employee needs when they need it. The leader and manager's job will be to figure out how to leverage the knowledge the employee has to maximise their effectiveness and make the company more successful.

This new management attitude extends beyond the organisation. Managers and leaders must take the same approach with associations, professional organisations and even some of their competitors. They must be 'willing to exchange information, ideas and products with allies and competitors [and teams] to stay on top of new developments and opportunities' [2]. Sharing is a two-way street. While secrecy may keep your knowledge hidden, it does not allow you to glimpse others' knowledge or ideas. Sharing some information, at professional conferences, seminars and university symposiums, helps the company understand more of the environment and better prepare for future surprises, resistance or changes.

It is interesting that many organisations in our current industry do not understand this sharing and interchange phenomenon. Many encourage, reluctantly, the participation of their employees in outside societies and organisations where they make presentations at conferences, but little is done with this information and often little acceptance by existing management is shown. Gathering the data accumulated by the participating employee can make a strong and helpful contribution to the existing products and services offered by a company as the employee is attempting to explain his/her understanding to others who are unfamiliar with the product or services. What is being suggested here is that management needs to be more receptive to participation and returning data than it currently is; today it is just being tolerated. It should be reviewed by the leadership upon return of the participant and gathered for impact on their current product and services. Discussion should also be allowed by others who may see a need for change or adjustment to improve the capability of the company.

A company often has one individual who is participating in these activities and who may excel over the other society participants. This accomplishment should also be encouraged as it often lends to the superiority of the company's representation and reputation. There is also the potential that ideas and concepts that are shared with the society can have an effect on shaping the industry, its products or its services in the long run. This becomes a plus for the company for whom that employee works from several positions. First, the shaping of the industry can be of advantage in the products the company produces; and second, the leadership development that is taking place can be provided from nowhere else in the current industry. This individual is actually leading several others in other industrial complexes to do things that the proprietary company already does well. This is a leg up and a business advantage of the first order for the company making the delivery. Yet, if you take a look around your company you will find those people working twice as hard to convince others in top management of the value of their society work. This is especially true for the specific managers who oversee the employee. We are in a global economy and we must view involvement as a resource and contribution to a company's global reputation.

3.2.1 Process and teams

New leaders today recognise that to be successful they must do unconventional things that might be indicated or dictated by the business environment. Today, changes within the company that improve results and products must be managed and encouraged. Acceptance of change is required and must be supported by the company's leaders. However, a structure or means of managing the change also needs to be implemented. If leaders and managers are to delegate responsibility for productive processes to their teams, the same method of managing and documenting the changes must also be managed and shared with all stakeholders.

The results of a change review process must be integrated into the system, and the means for that integration must become company policy as part of its standard operations. Leaders must give up their micro-management of the day-to-day operations to support the process improvements identified and necessary for improved operations, services or products. The improvement of the processes belongs to the specific production team, the established and accepted processes to the company. Therefore, ensuring that process changes go through an effective and efficient process review operation must be the responsibility of the company's CCB and its respective leaders.

The evolution from the 'hero' to a fully functioning team- or process-oriented system requires a great deal of work on behalf of the key process holders. Each has to give of their ideas, time and knowledge to create processes that will allow the organisation to prosper and improve. Change as it occurs in the company and especially the Change Review Process must be accepted and become the 'rule' that everyone can live with. The first step is to establish an Integrated Product Development Team (IPDT) or CCB that has the authority and responsibility to examine the existing processes in the immediate value streams. This may occur at the encouragement of a single or multiple member(s) of a production team. These reviews must be pertinent to product development. Following its establishment, CCB must become the authority for approval and dissemination of results as informational change and direction for those involved with the specific process. With all the data available, the CCB must first define accepted processes and get the stakeholders to begin their value stream reviews. Change proposals must be reviewed by all involved to ensure that the process owners are able to apply the best and most efficient processes in improving product output.

Keep in mind that the complete development process considers all the steps necessary to achieve the desired goals or lifecycle state of the product. The process is a disciplined approach for developing the product as well for managing the overall development. The life-cycle usually describes the product process in terms of the product states and development phases. The lifecycle perspective defines the states that a product reaches as it matures over its useful life. Understanding the phases at different points is critical to measuring the developmental progress towards its objectives and ensuring that the stakeholder commitment is maintained, identified and resolves the communication or integration issues in the development phases.

- Tasks are not easy
  - Good chance for completion
- Team location is ideal
  - Reduced interruptions
  - Good facilities
- Tasks have clear goals to succeed
- Immediate feedback is provided
  - Adjustments are accommodated

Chart 3.7   Successful Teams [Source: From Csikszentmihalyi, M. Flow: The Psychology of Optimal Experience. New York: Harper Perennial; 1990]. Note: See Appendix section for full page image

The teams must work together to manage and document process improvements. This is essential to removing the 'hero' and placing the current processes at the centre of the business. Each process team has the authority and responsibility to examine the existing processes in their immediate value streams pertinent to product development. As an example, a CCB can be established to manage changes in complex engineering or manufacturing operations. With all the data available, the CCB must coordinate with each team to define desired processes and their changes, then work with teams to begin developing the value streams. The effected teams then recommend and document change proposals to ensure that they apply the best and most efficient processes to improve product output. These improvements must be documented.

With the changing demographics of the American work force, the scarce resources today are knowledge, entrepreneurship, and more generally human capital. This shift from an emphasis from financial capital to human capital has significant implications for leadership philosophies. Strategy, structure, and systems thinking will be replaced with purpose, process, and people thinking – getting people to help define and then align with purpose, developing the processes to accomplish the purpose, and then attracting and maintaining the people to push the processes. [22]

There are five phases to process engineering as developed by the Software Productivity Consortium and their resulting courses and manuals [23]:

- Phase one is to establish a plan for the overall effort. It must involve all of the managers, leaders and employees in the company and all of the product teams. The phase will require a considerable amount of time from all the participants. The plan should look at how the process engineering phases can be established most effectively with the least amount of effort on the overall function, but the phase must be communicated such that everyone in the company will be involved in some form or another, their assignments to follow the plan development.

- Phase two is to capture all the processes being used, while at the same time gathering input from the product and service teams as to how the processes can be improved and more efficiency added through the use of value stream analysis.
- Phase three identifies the lifecycle for each product or service and establishes the models which document the various phases through which the product transitions.
- Phase four establishes a standard process architecture that allows each process to be supported. This standard process must be constant for all the projects, products and services. The optimal architecture meets all the users' expectations, needs and requirements. It allows the process to handle all expected and unexpected growth and is cost effective.
- Phase five defines the standard process so that each process element can operate by definition, be a process asset in the data library and become a part of the database that supports the organisation's process.

3.3 Leadership in a capability maturity model

Capability is really the name of the game in today's work environments. A company cannot get a job completed satisfactorily without having employees who have, as a matter of being, the ability to work in an area and know from experience and training that they are doing everything right, correct and acceptable according to quality standards and plans. When a manager hires an employee, based on the interview, the experience provided in the resumé and the background provided by the references, there is the expectation that the job can be done and will be done according to the accepted process supported by that manager. Without the expectation of capability, a manager is shooting in the dark with the person they have hired to do the job. Therefore it is necessary for the manager to know what capability will be expected to execute the job or role one is to be filled. This capability must be based on the specific competency assigned or credited to the team.

In order to assure the company (and its customers) that the capability is there, many organisations are looking to frameworks known as Capability Maturity Models (CMMs) as standards to measure the range of strengths in that organisation. Measurement of these looks at the skills, abilities, knowledge, processes, methods, experience and tools used in the production of a product or service. The CMM provides such a measure for those capabilities and tells the organisation where they must improve and correct their capability gaps.

The first efforts at a Capability Maturity Model, Version 1.1 (CMM 1.1) were exercised by the Software Engineering Institute (SEI), in the early and mid-1990s. The exercises aimed to provide the armed services (mainly the Air Force) with a mechanism that would assure quality software, reduce the cost of the product to the government and improve the producer's ability to make repeatable and quality software the first time, every time. An effective self-management system was developed that would assure an operational programme was being handled efficiently and effectively throughout the products lifecycle. What was found was that the management principles worked for the software establishment and the return on that investment exceeded the cost to establish it. The SEI established a mechanism to develop operational and company maturity and put established fundamentals that supported the development of quality products that could be replicated over and over again. This is what a facilitating client and company wants: the ability to repeat the productive process over and over effectively with a positive quality result.

CMM 1.1, whose use this textbook encourages, has five levels. These are illustrated in Charts 2.1 and 3.8. By definition, following the levels and developing an organisation that can demonstrate and use the levels up to level 5 will provide an improvement that reduces risk and increases the productivity and quality of the product.

	Level	Focus	Key Process Areas	Result
Improve	**Optimising** 5	Continuous Improvement	Process Change Management Technology Change Management Defect Prevention	Productivity & Quality
Control	**Managed** 4	Product and Process Quality	Quality Management Quantitative Process Management	
Defined	**Defined** 3	Engineering Process	Organisation Process Focus Organisation Process Definition Peer Reviews Training Programme Inter-group Coordination Product Engineering Integrated Management	
	Repeatable 2	Project Management	Requirements Management Project Planning Project Tracking & Oversight Subcontract Management Quality Assurance Configuration Management	R I S K
	Initial 1	Heroes		

Improvement initiatives must increase market share and/or profitability in order to have business value!

Version 1.1
C-SEI/CMU

Chart 3.8   *A Process Management Model ... the Capability Maturity Model*
[Source: Image adapted with permission from Zubrow, D. 'Putting "M" in the Model: Measurement in CMMI'. Carnegie Mellon University, 2007].
Note: See Appendix section for full page image

Level one is often called the initial stage. To emphasise the error many companies use by operating at this level, many still call it the 'hero' stage. It is characterised as an ad hoc process and product orientation and can only be described as chaotic. Processes, methods and tool use are not documented and are communicated to the worker in a disorganised manner usually at the moment of need. Successes derived from operations at this level are often attributed to a single individual whose efforts are considered heroic but are often not repeatable. Risk is usually high at this stage, with risk mitigation on a catch-as-catch-can basis.

Level two is called the repeatable stage. Project or programme management fundamentals are used where processes, methods and tools are planned and tracked for cost, schedule and functionality. Documentation is in place that focuses on the same process, methods, tools and abilities to repeat past successes and a record that can be followed to support that ability. This is attributed to the control of the six key processes in the stage. Risk goes down while product quality improves. The six key processes are:

1. requirements management,
2. configuration management,
3. quality assurance,
4. sub-contractor management,
5. project planning and
6. project tracking and oversight.

These are key items, not only key processes. Using them allows the programme manager or leader of the project group to be functional in their operations, while maintaining an ability to repeat what they have done over and over again with very little variance in those applied functions.

Level three is called the defined stage. Processes, methods and tools for both management and engineering activities are documented, standardised and integrated into the organisations standard operations. The operations and processes used by the company in its products are more consistent across the organisation where support is provided to the products and services by process improvement functions and training. Again the risk is greatly lowered and the product quality increases. Training is probably one of the most important key processes of level three. The training competency structure is illustrated in Chart 3.9, which illustrates how the capabilities for a role are built from the knowledge gleaned from the stakeholders. Developing the 'body of knowledge' for each role makes the team that much more able to get the work completed effectively. It is also suggested that a project leader have the key process of training in play even if they are not operating at the level three stage in their organisation. Operating with the 'training' key process and all the requirements enables the leader and their organisation to operate more effectively and efficiently.

Besides the key process of training, the other key processes for stage three that are included in CMM 1.1 are:

1. organisation process focus,
2. organisation process definition,
3. peer review,
4. inter-group coordination,
5. product engineering and
6. integrated management.

Each key process has its own values and works well in assuring the productivity of the organisation and the product groups.

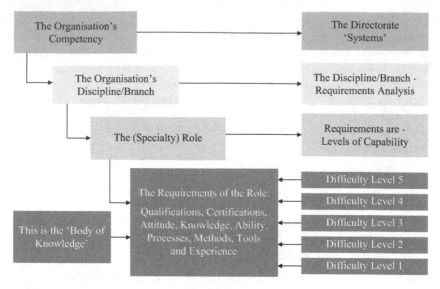

Chart 3.9   The Competency Structure. Note: See Appendix section for full page image

Level four is called the managed stage. Measures are more detailed for all the processes, methods and tools where product quality data is collected and used to manage the product development. Statistical process control techniques are used, along with the qualitative method for managing the project, product or service. The key process areas focus on the qualitative appreciation for both the process and the work product. There is also an emphasis at this level on cooperation and inter-communication between the parties. Being able to know what others are doing and what changes are being applied makes the process that much more applicable. This focus reduces the risk again and product quality increases. The key processes of level four are: quality management and quantitative process management.

Level five is called the optimising stage. Optimisation is enabled by the quantitative feedback from the continuous process improvement steps taken. These allow innovation and technological improvements to be applied to the project, product or service processes. The organisation is expected to implement continual, measurable process improvement across the company. The key processes are: process change management, technology change management and defect prevention.

CMM 1.1 is only one example of the systems that can be utilised by companies according to their comfort zone and experience in the field of using models to improve their operations. Some others that can be looked at are:

- ISO/IEC 12207,
- IEEE/EIA 12207,
- MIL-STD-498,
- J-STD-016,
- SYSTEMS ENGINEERING CMM,

- EIA/IS 731,
- EIA 632,
- IEEE 1220,
- CMMI and
- ISO 9000.

Each has its own special interest and ability to improve organisational capability. As the reader can see, there are an awful lot of different models out there. The best bet is to pick one that really works for you. However, the author feels that thanks to its focus on the project management phases and its resultant support for meaningful results for a product or service, CMM 1.1 is the most applicable to the most common engineering fields and community [24].

3.4 Programme and project management fundamentals

Key processes follow a general rule; they should always start with the customer. The first process at level two of the CMM that must be ensured is defining the customer requirements for the product. The company must manage the translation of customer requirements to actual product specifications.

The product specifications baseline, as dictated in the original requirements, must also be maintained through a configuration management system. At the same time configuration management must resist unnecessary changes while allowing only essential engineering and customer-driven changes as they occur and are

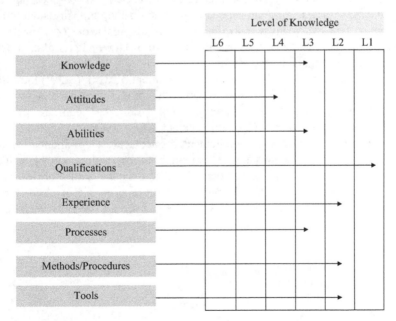

Chart 3.10   The Body of Knowledge. Note: See Appendix section for full page image

approved by the CCB. The customer and product quality must not suffer as the product or service is developed and produced. This is what we saw in the *Columbia* example given in Chapter 1, where one of the most important requirements – safety – was given a non-important role in the programme.

Several components are involved in analysing project, product or service requirements. I have identified nine. They are given here in no specific order, but they are all necessary if you are to be committed to a full analysis:

1. contractual requirements,
2. non-contractual requirements,
3. requirements prioritisation,
4. identification of any unclear requirements,
5. allocation and specifications of derived requirements,
6. requirements flow-down as they fit the work breakdown structure (WBS),
7. traceability and audit requirements,
8. accountability requirements for responsible parties and
9. the project manager's responsibility to the project, product or service.

Contractual requirements can be brought to light. A letter of intent might be submitted to the team leaders' attention and asked to be given some review and assessment. This may come from the marketing or management levels. The letter of intent is only an interpretation of the customers' questioning the ability of the organisation and implies no contractual obligation from either party. However, it does give an indication of interest that can be researched by the engineering team if they feel that there is a genuine need or interest on the company's and customer's behalf. It provides advanced information of potential work to the seller for planning and assessment. This research might be followed by a letter of contract or acceptance by management that is used for the start-up of a project. This is a contractual obligation in a letter format or a statement of work (SOW) that limits the scope and spending of the company. Often this provides only a small amount of funding to encourage the company's organisations to negotiate with the potential client, if there is one. Once the contract is negotiated and signed by both parties or given the go ahead by management, the actual work commences. It must be noted that on some occasions a memorandum of understanding may be included to focus on things that may have been overlooked in the actual obligation or contract. This approach would only be used if there is an actual client for the service to be rendered. If a contract is required, it would typically include the following as listed:

1. statement of work,
2. work breakdown structure,
3. performance statement,
4. contract deliverables,
5. acceptance criteria,
6. data or documentation requirements,
7. schedules,

8.   referenced standards and processes,
9.   payment provisions and
10. union labor contracts.

The non-contractual requirements would include:

1.   company policies and procedures,
2.   individual customer priorities,
3.   OSHA and ISO regulations,
4.   total quality management,
5.   marketing priorities,
6.   government regulatory requirements,
7.   marketing window and
8.   competitive analysis.

Requirements prioritisation puts the company's leaders in the position of understanding each of the necessary specifications as recorded in their directions and their relative priority to the overall project, product or service. Configuration management procedures help leaders determine the placement of parts and the priorities as well as the construction of the WBS.

Identification of unclear requirements means that the leader or manager is aware of the use of certain terms, and they try to avoid them. Some of these unclear terms are:

1.   'user friendly',
2.   'simple interface',
3.   'easy to learn',
4.   'easy to use',
5.   'in accordance with the best modern standard practice' and
6.   'highest quality workmanship'.

What do these terms mean? No one really knows the answer to that question, which requires the leader to sort out the ambiguous terms and get real clarification for the processes to continue. Unclear requirements can cause a lot of problems for the project if they are not clarified up front. As one might assume, the reason this is brought forth is that ambiguous terms must be identified and clarified for the company's best results. If this is not done, one might find the project team chasing an illusive requirement that might never be met.

Allocated and derived requirements are those that have been put in through the back door with the funding or by citing the use of a specific operation that can only be done in a way that requires tools unspecified, but by their very nature are derived. They are derived when they are expanded to a lower level of detail by a quantitative analysis. They are allocated when they are expanded to a lower level of detail using applicable past experience and managerial judgment. This is also known as 'requirements flow-down as they fit the work breakdown structure'. The allocated and derived flow-down naturally cycle to the traceability and audit requirements – which are precise. The data helps make sure things are happening the way they are supposed to. That is, are the correct and appropriate processes being used to develop the product or service?

Leaders and managers of projects involved with contracts and written requirements should be aware that there are items that have to be verified. The accountability requirement ensures that management or leadership has incorporated all the components into the product and they are able to verify that by:

1. test,
2. inspection,
3. demonstration and
4. analysis.

The project manager's responsibility to the project or service is to ensure that all the requirements are clearly defined and documented if there is an agreement involved. It is recommended that all the parties to the project sign the agreements, with the executor and the user or requestor in the client producer position. This agreement ensures that all changes are managed without affecting the product, and that all involved are aware of the 'constructive changes' that might result. The leader or project manager ensures that the requirements flow-down to the lowest level is appropriate and uses a traceability system to manage each of the traceable items. The leader verifies the conformance of the items as the design evolves, conducting frequent design reviews and maintaining a compliance matrix. In retrospect the project leader is responsible for rigorously reviewing and controlling the changing requirements to ensure that the change control system, as established, is working correctly and is providing the appropriate feedback to management and leadership at all levels of the company and the project.

What does it take to lead an organisation in today's complex environment? Successful leaders do a good job of estimating, predicting and guessing about the future. Frequently, they are correct or close enough. But it's always a guessing game. Realistically, the hero's environment is basically unknowable, uncertain, nonlinear, complex and rapidly changing [1].

To improve, leaders must create an organisation that is more adaptable to their business environment. Creating a process orientation throughout the whole organisation provides the desired adaptable and flexible structure. Leaders must go beyond speeches and slogans and have a strong desire to understand their company's process flow. For this reason the author recommends using the planning process suggested in Chapter 2 and illustrated in Chart 2.3, the work breakdown structure.

For this reason the leader must be aware of the operational tools that they have available to them. These are the constructive requirements analysis, configuration management, quality management, subcontractor management, planning, oversight and tracking of the project or service.

> Leaders that lead by collaboration, compassion, communication, and values, not by planning, organising, directing, staffing, and controlling; are leaders capable of bringing (operational) energy and understanding to

the local challenges and at the same time integrating those (organisational) local actions with the organisation's purpose and direction. [19]

Most companies are driven by the desire for results and profit through production of a product or delivery of services. Rightly so, as productivity, resulting in profit is essential to its survival. However, most companies, once they set up a department or procedure, ignore the established process and proceed to cajole, harass and badger the employees to increase their productivity. The official process used by the employee to produce the product or service is ignored. And once again only results count. Most employees get their work done, ignoring the official process, procedures and department silos. 'The informal networks, the practical decisions and actions, and the common sense in doing a job end up driving the day-to-day operations in most organisations' [5].

Leaders must balance the teams' need for autonomy and flexibility with the needs of the organisation for control of resources. In a very dynamic work environment decisions need to be made as close to the customer or company/environment interface as possible. This primary contact is the employee dealing directly with the customer.

Leadership and the management need to manage the resources used by these decision makers. All members want to maximise their self-actualisation while fully moving towards the organisations strategic goals. Those opposed to this are those who deep down don't trust others to work in the right direction.

Organisations function in many ways to meet their goals and objectives leading to a quality product or service. 'The challenge of this new leadership is to create, maintain and nurture their organisation so that it creates and makes the best use of knowledge to achieve sustainable competitive advantage .... Collaboration is such an important part of creating the right environment and leveraging knowledge' (see Chart 3.13 [9]).

- Essential

- Doable

- Affordable

- Describable

Chart 3.11 Project Requirement Management Essentials. Note: See Appendix section for full page image

Can we establish an approach that ensures the transfer of leadership over time as is required, or do we stumble, retrain, restart and disassemble/reassemble when things change due to major upheaval? Knowledge can no longer be transferred only in the traditional method through books or in a classroom. Instead, through on the job training team members ensure that each of them is able to do each other's activities and act interchangeably. Transition of work activities and knowledge need to be dynamic, frequent, and result with a minimum amount of disruption where the function continues and the work gets done.

'Complex adaptive systems cannot be controlled, they can only be nurtured. Control stifles creativity, minimises interactions, and only works under stable situations. It is not possible to control a worker's thinking, feeling, creativity or trust' [8]. There is now more information than an individual can comprehend, more change than a manager can control. Organisations that don't adapt will not be able to operate in this new environment. 'Only knowledge can provide the understanding needed to deal with this complexity. Such a milieu demands a different paradigm and new rules and roles for leaders and managers' [9].

An organisation must continuously verify that it is satisfying customer requirements. All key processes must be examined, using this rule; they always start with the customer and work backward.

> As society becomes more and more complex and uncertain it will be harder and harder for individuals and organisations to control their external environment. What is happening now, and will continue to happen in the future is that the successful organisations will be those who have developed the capacity to co-evolve in an ecological sense with their external environments through mutual interaction, internal adaptability, and rapid response. These organisations will develop a strategy, structure, culture, and overall health level that permits them to act intelligently, creating, leveraging, and applying knowledge in a manner that leads the to overcome environmental threats and take advantage of opportunities. [25]

Before our own processes are examined, we must first clarify customer requirements for the product. Some may object that there is no one customer, as their product is sold to many. But, there is the ideal or typical customer. With research methodology, they can be queried and evaluated. Process experts can then translate these ideal customer wants into customer requirements. Engineers or product experts then will convert those requirements into product specifications.

When looking at the product requirements, the leader of an organisation must ensure that the specifications provided are correct. To do this might require many people from the various silos of the organisation. This was the reason for concurrent engineering and management and the development of the integrated process project teams. Its processes integrate the knowledge of many to analyse the required specifications. The requirements must be clear, unambiguous and consistent with the processes that allow for the development of project. The leader must assure the company's leaders that the 'what' is stated clearly, rather than how it will be done. It must be testable and independently verifiable under all conditions, especially those in the company itself.

Knowing that the specifications or requirements are appropriate, the company processes can now be put to work in manufacturing or providing the product or service. The process experts will work from product specifications back to the original raw materials. They will create a production processes to manufacture the product or provide the service to the customer. Once the processes are developed and in operation, the quality of the product or service is maintained by using a configuration management system to benchmark product specifications. The customer and quality requirements must not suffer as the product is developed and produced.

This is the value of the WBS and its continued assessment. The WBS aids in the development and scheduling of the tasks to be completed. Reading the WBS and all the tasks that must be done allows the leader to assess the resources needed and to assign responsibility. From this most useful tool comes the ability to determine the products or deliverables that must be developed and to establish the project work authorisation agreements that must be developed with the subcontractors and all other contributors to the project. As shown in Chart 2.3, the cost accounting and budget can now be developed. To list only a few, the specs can now be determined for the deliverables that the tasks call for and for the subcontractors as well.

Once the first cut at the WBS is made it is imperative that a baseline be established as the starting point for the project. This is of course after the schedule has been balanced for all material and resource requirements. Now it is time to fix the dates for all of the milestones; it may be necessary to adjust the resources and cost as time and schedule often cause these financial items to vary. Preparation of the project, product or deliverables list may require corrections to the WBS, its milestones and the task directory, which may also require additional milestones. Now that the leadership knows what the baseline looks like, it is time to look at where these project work assignments are going to go, whether to subcontractors or internally to the company's staff. Project work authorisation agreements must now be dealt with and assessed for completion as meeting the requirements set for these deliverables. Company policy must also be considered regarding how these are to be let out. Now we can capture the first baseline schedule and items with the related data into the permanent configuration management tool as the established record.

It is now time to publish this plan and make sure that all who are involved or responsible are included in this publication. At this point the WBS and its baselined plan are open for re-planning, and justifiably so. The ideas and suggestions of the personnel involved should be used to correct and change what must be changed. However, this should be controlled by the CCB or group that has been established in the company. Critical path software can show where the changes need to be established and will have the greatest effect. The original project goals must be reviewed in light of the suggested changes so that the affected teams will not be changing the project, its product or service and the interrelated deliverables. Do not re-baseline unless the requirements have changed as a result of upper management's reevaluation and negotiations, as this will shorten the critical path. Be sure to document all agreements and produce new versions of the task dictionary items, charts and so on. Make sure the changes are integrated into the WBS numbering system. Once a plan or project has been baselined, all corrective actions must be taken to recover the original plan itself or the accepted facsimile. New baselines should be version controlled to indicate the baseline change and dated to indicate when this was approved by the CCB.

Re-baselining should only occur following a requirements or configuration change; all other re-planning should be done to correct problems that have been discovered in the original planning process. Again, documentation and data collection in the configuration management system is of the highest priority.

3.5   The importance of dealing with the culture

While it is important to establish the baseline structures of requirements and configuration, it is also important to be cognisant of the culture within which this product is being built and the culture to which the product is to be used. Leadership needs to develop and maintain its own culture, emphasising quality and cost effectiveness while maintaining a focus on customer requirements. This chapter will sum up the impact of the culture on the building of a product, the deliverables, the products' customer culture and the resulting long-term success of the product.

> When conventional wisdom becomes too beholden to the past, however, an organisation's culture grows stale. [26]
>
> The future is truly unknowable and therefore we must learn to live and deal with uncertainty, surprise, paradox, and complexity. [8]
>
> Culture by definition is elusive, intangible, implicit, and taken for granted. But, every organisation develops a core set of assumptions, understandings, and implicit rules that govern day-to-day behavior in the workplace. [27]
>
> Organisational culture is a common set of perceptions held by the organisation's members; a system of shared meanings. It is a set of key characteristics that the organisation values. [28]

Inside an organisation, especially if you have been employed for some time, these perceptions become subconscious and everyone takes them for granted. Even someone inside, they may not know these values, if asked. Core values are the primary or dominant values that are accepted throughout the organisation [29]. The best way to see common perceptions is to compare one organisation with another. Look for culture characteristics in each and the comparisons will become obvious. 'Culture is the social glue that helps hold the organisation together' [27]. Social relations give everyone connected to the company a bond. Whether a company picnic, Friday afternoon sales and BBQ, speeches, open houses, etc. all can be used to communicate what behaviour is acceptable and desirable for the stakeholders or employees. On the other hand, people who are hired with the wrong values will, after a relatively short period of one to six months, either adapt and internalise the company values or leave. This is why we must look at the age culture differences that Marilyn Moats Kennedy has pointed out (see Chart 3.12) [17].

Transgressions of the rules on the part of high-level executives or front-line employees results in universal disapproval and powerful penalties. Conformity to the rules becomes the primary basis for reward and upward mobility. [27]

Characteristics Also known as: Born Between:	Pre-Boomers, Veterans, Silent Generation, Seniors (1922–1943)	Boomers, Me Generation, Sandwich Gen (1943–1960)	Generation X, Busters, Cuspers (1960–1980)	Millenials, Echos, Nexters, Generation Y (1980–2000)
Work Habits:	- Follow Tradition - Status Quo - Obedience over Individualism - Advancement Through Hierarchy - Sense of Duty & Honour - Natural Leaders	- Value of Personal Growth - Wants to be Involved - Team Orientation - Value Company Commitment & Loyalty - Sacrifice for Success - Uncomfortable with Conflict	- Entrepreneurial - Independent - Thrives on Diversity - Desires High Level of Responsibility - Constantly Looking for Creative outlets - Quickly Moves on if Employer Fails to meet Needs - Impatient	- 24/Seven - Capacity for Multitasking - Global Connections - Competitive - Civic Minded - Diverse - Desire for Structure - Goal & Achievement Orientation

Chart 3.12   Culture Differences. [Source: From Kennedy, M.M. 'Career Strategies'. Presented at ASEE College Industry Education Conference, ASEE, 2007. Available at www.moatskennedy.com]. Note: See Appendix section for full page image

Collaboration requires a close, open, and trusting relationship where each party contributes their capability and works with others to align and integrate the efforts of all. Leaders use collaborative relationships and interactions to share understanding, get the work done, and guide development of their coworkers. It is through a collaborative approach to relationships that leaders earn their leadership rights while at the same time serving the knowledge workers. [30]

The leader who understands age culture differences and is able to work with them to integrate these differences into the company culture is using their organisational skills to the best advantage and is going to get the best work out of the employee/stakeholder. While the company culture takes precedence, differences of age culture need to be considered when working with the employee.

3.6   Seven characteristics of organisational culture

Every leader must be able to recognise the key components of their organisation's culture as well as how to change or strengthen it. Robbins [31] identified seven characteristics of an organisation's culture.

The first is the amount of innovation and risk taking tolerated. Does your company encourage risk taking or punish it? FedEx's reputation was established as a company that encouraged innovation and the bypassing of rules, if it meant pleasing the customer. However, if you are a manufacturing company, the production manager may be punished for changing the established production steps without approval by the appropriate group. Consistency and dependability in meeting production schedules would be rewarded instead.

> Maintaining optimum complexity is the ability to generate ideas and actions (increase its internal variety) that are creative and innovative and allow the organisation to make use of, harness, or overcome comparable complexity in the environment. ... This generation of additional variety is a cost in terms of time and manpower that has to be traded off against direct application of energy and mental capabilities toward customer needs. [32]

This is why risk management or risk assessment is taken up at the very start of a project or service activity. Risk or opportunity, whatever you wish to call it, begins at the very start of the activity. The WBS chart is a decision-making chart with the thought-provoking questions that need to be considered (see Chart 3.13). Notice that the pre-planning steps include the risk management questions.

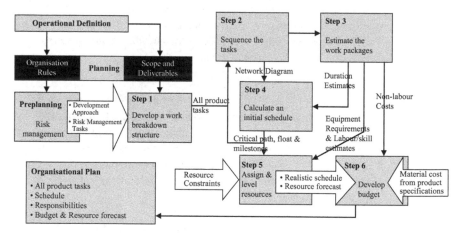

Chart 3.13   *Developing the Work Breakdown Structure (WBS): Decision Making [Source: Reprinted with permission from Verzuh, E. The Fast Forward MBA. 2nd edn. New York: John Wiley and Sons; 2005]. Note: See Appendix section for full page image*

It is important at the start to identify the risks and opportunities before the WBS is developed. Questions such as: 'What could go wrong? What opportunities are available? These can be stated as 'If this ... then' statements and grouped by category. The evaluation should be made on both probability and impact, with prioritised outcomes. Next one must develop feasible actions to enhance opportunities and mitigate the risks. From there it becomes a cost assessment based on immediate or contingent actions.

Second, how much attention to detail should you encourage and reward? If you supervise a bank operation or electronics manufacturing company, then attention to detail is critical. However, if you lead a sales team, making customer contacts and closing sales are far more critical than making sure the details on order forms are accurate. Once the sale is completed the details can be handled by a detail-oriented person in the home office, not the salesman. There are two types of risks. One is known as development risk and the other is product risk. Development risk is the chance that the planned events will not occur as planned, they are often technical, related to schedule and cost and have other people as the main function causing the errors. Product risk is the chance that the system will fail or cause injury. They are risks that are inherent to the system's use and effects that result from the failure of its operation.

Third is your outcome orientation. Which is more critical – accomplishing the mission or following procedures accurately? If you are making sales, or pushing to meet construction deadlines where penalties for missing dates will be incurred, then results orientation is critical. However, if you lead an engineering operation, working for NASA or an operation with significant government oversight, then following documented procedures is far more critical than actually producing a product. And deadlines can be renegotiated. As impractical as it seems, the end product is inconsequential compared to completing the checklist properly. Many engineers never actually see the company product; they are more focused on accomplishing proper procedures, like six sigma quality, cost centre management or configuration management. For this reason I bring forth the issue of the two types of opportunity: they have the potential to change the company and its approach. There are strategic opportunities and tactical opportunities. Strategic opportunities look at the new market potential where a new product might be introduced, or a new approach. This is where an increase or decrease in pricing might be introduced. Tactical opportunities shorten the schedules and reduce cost. They can also mean improvements to the process that eliminate unnecessary activities and leverage workers' time to effective applications.

Fourth, is the leadership or management most concerned with the impact of their decisions on company profits or the wellbeing of the employees? Does the company leadership see its employees as expenses, figured into the profit formula, or as assets to be developed for the long run? Are they willing to sacrifice short-term profit for the long-term development of their employees, to increase loyalty, which results in long-term company growth? Keep in mind the age culture and how this will affect them and the management's thinking about the job. One of the biggest problems most government contractors face is that they are busy cutting costs by eliminating people from the mix and then do not take into account the changes to the tasks and processes that must be set in motion. Often the manufacturing or engineering group of the prime contractor ends up eating the extra effort required on behalf of the existing personnel when 'expendable' personnel are cut to shave costs.

The fifth characteristic is team orientation. Who is rewarded for performance results – the whole team or an individual? If the individual is rewarded, then a competitive environment exists and teamwork will be shallow or non-existent. This is why it is so important for the team to be developed with people who can and do work together well. Often company politics is played to populate a team, with the result being only a few working and the others riding on the productive workers' deeds to garner the rewards provided by the company. Team development is of the utmost importance for proper team orientation. It is fundamentally important for the project leader to have a good set of role descriptions for the types of members that they wish to have on the team. If company management is dictating the membership, the project leader is already in a losing position. It is imperative for the leader to insist that they be given the responsibility to hire, fire and recruit the people they need to have. This should be based on the development of the roles and responsibilities identified in the WBS. A company that does not develop its known competencies, disciplines and specialty roles is a company that will eventually fail or will do a poor job of completing its tasks. This will be evidenced by the high cost of production, excess employees and overruns on its budgets. Coaching, training and mentoring probably does not exist in this company, or their capabilities are grossly overlooked.

Combine this with the sixth characteristic: what is the atmosphere like? Do employees work together or are they competitive, aggressive, secretive and cut throat? It goes without saying that the atmosphere in the company just described is one of deep despair and lots of looking over one's shoulders. No one trusts anyone, and if you are not sure what it is that you are supposed to be doing, then you certainly don't know what the next person is supposed to be doing. For this reason it is important that the competencies be identified, what the appropriate disciplines are corresponding to those competencies and that the specialty roles have been determined. When roles have been assigned and everyone knows what they should be doing and what others have been instructed to do there will be no confusion in the ranks. It is especially rewarding when someone knows that they have been picked because of their expertise in some areas and that training, mentoring and coaching are available from the leader as necessary. This establishes a very positive atmosphere where everyone working in the environment appreciates their position and the skills they are able to demonstrate and learn.

Seventh and finally, what is the long term perspective? Is management focused on the long-term growth of its people and the company or is it concerned with keeping things as they have always been [31]? Knowing what the long term has to offer is a huge plus to those working on the teams. If they know what the plan is for the project, product or service, they can determine what they are going to be doing, for what length of time and what the potential is for the future.

What is the role of culture? As consumers we can argue the differences between Coke and Pepsi, between Ford and Chevy, or between Microsoft and Apple. We can see qualitative differences in any of these corporations. Cultural differences exist between all companies. We as humans can sense the difference and react differently when in the presence of those entities. Those who travel between companies, such as salesmen or consultants, can tell the difference, whereas those working for one company have become desensitised to cultural nuances. Culture is a unique sense of what an organisation like a company stands for. This uniqueness gives members a means of identification that some know and others simply assume. Members become very loyal to their company through this identity. Don't try to give a Pepsi drink to a loyal Coke employee. That loyalty creates a sense of commitment to the company. Loyal employees want the company to succeed and will often do what-ever necessary to get that message across. For the future of the company, 'culture is the social glue that helps hold the organisation together' [27]. Good leaders use it to develop and guide their employees.

- Early Identification of Problems/Risks
- Work Around Solutions
- Cost Avoidance
- Schedule Slip Avoidance
- Improved & Effective Planning
- Application of a Professional Way of Doing Business

Chart 3.14   Proactive Approach to Process Leadership. Note: See Appendix section for full page image

However, culture has its price. 'Culture is a liability when the shared values are not in agreement with those that will further the organisation's effectiveness' [33]. In rapidly changing environments, tradition bound company like Sears or Eastern Airlines were not able to change fast enough. The culture must not only support a quality product and customer service, but must instill a competitive company spirit that encourages its survival under all conditions.

'Senior leaders are expected to articulate and represent the culture. Managers and leaders are expected to help translate that culture through leadership initiatives. Employees at all levels below the senior leaders are expected to attribute meaning to various actions' [34]. The culture must encourage employees at all levels to maximise the effectiveness of the company's internal processes, to emphasise product and service quality and to be responsive to customer needs. If you don't know what your role is within that structure, where its competencies are changing and the disciplines are yet untrained, the responsiveness to the customer is non-existent.

However, simple descriptions can backfire if misused by the leadership or upper management. 'When a company like GE preaches 'lean and mean' as a culture, or a company like Enron preaches honesty as a value, employees are likely to see a very different meaning when they discover that the CEO (and others in upper management) were paid lavishly or the senior executive staff was dumping stock while urging employees to buy more' [34].Upper management might get away with being two-faced about company values for a short period. But their true values will eventually be discovered and employees will hold them accountable. The bottom line is that leadership needs to develop and maintain a culture which emphasises quality and cost effectiveness while maintaining focus on customer requirements that support the appropriate development of valued products or services.

Management must build trust in their established teams. This can be done through the leader developing their own team and not having one given to them by management. Traditionally, an employee has to earn the trust of their leader through many actions. Apple polisher, brown-nose, teacher's pet – these are all names for people who did whatever the leader or manager wanted in order to win favour. Today's leader no longer have the time to build trust and loyal followers. They must initiate the trust of their team members from the very beginning. Managers and leaders must take actions that put trust in the team, based first on their own choice of who will participate and who will not, until one of the stakeholders earns their distrust. Change occurs too fast to allow trust to be developed in the traditional sense [10].

> From the collaborative leadership perspective a dynamic balance arises from the issue of how much control versus freedom the teams and knowledge workers should have. [10]

> In a dynamic environment single individuals are not smart enough to be able to interpret all the needs of the market place at the local or international levels. While recognising that hierarchy, responsibility, and some level of authority is essential to any organisation, the issue of the self-organisation of teams and the degree of freedom and empowerment of individual knowledge workers and their teams is of constant concern (to all involved). [34]

Companies that have successfully built a process-focused culture will reap the benefits. 'The success of the Toyota Production System was grounded on the belief that if you are observant of the process, you are in a better position to make improvements' [2].Unfortunately, 'there are too many leaders looking for a quick fix, the new tool and method, the next wave or fanciful idea. ... Executives ... thinking those things will resolve their problems, when they have little understanding of their business and how it works. They are far removed from their processes and customers' [2].

> If you stand in a (Toyota) factory and use your perceptual intelligence (direct observation), you will observe two types of phenomena – movement

and connections. The primary movement you see is the physical progression of raw material becoming finished goods. But, without the discipline of looking, we miss an enormous amount of the waste in the movement. To see connections you must stop the flow and observe one movement in the process. Take a workstation or machine and look at the surrounding physical components that directly affect that operation. Then push that out to see the next level of connections, like maintenance, standards, scheduling, training, lighting, measure, etc. [2]

Detail process focus is how you wring out every drop of waste. It's how inefficiencies, delays and bottlenecks become obvious. Only if management continuously reinforces the emphasis on process orientation will the company be truly competitive. This is why it is so important to be the leader who trains, coaches and mentors their employees. This process lends to the employee understanding their relationship to the variances, how to deal with the customer and how to assure that the subcontractor is doing what they are supposed to do.

1. Tolerated Innovation & Risk Taking
2. Attention to Detail & Reward System
3. Outcome Orientation
4. Impact of Leadership Decisions
5. Team Orientation
6. Company Atmosphere
7. Long-Term Perspective

Chart 3.15   Seven Characteristics of Organisational Culture. Note: See Appendix section for full page image

Everyone benefits from a culture that does not seek to blame and that is focused on using data to make improvements. Setting targets helps clarify direction. ... The use of value stream mapping (and analysis) to establish the current state is essential'. [2]

Practical techniques for seeing the core process include visual management and employee engagement. Kanban and Five 'S' were developed in response to the task of improving process visibility. Unfortunately, when introduced in the West, Five 'S' was often called housekeeping. Cleanliness and orderliness which are only sub-goals; the main purpose of Five 'S' is to promote process visibility that is to make kaizen opportunities instantly obvious. [2]

Management (and leadership) must emphasise the internalisation of procedures that continuously encourage the employees to improve their specific work areas and flow. This is again the perfect opportunity for teaching, mentoring and coaching.

'Whatever an enterprise does internally and externally needs to be improved systematically and continuously: product and service, production processes, marketing, service, technology, training and development of people, using information' [35]. All levels of the organisation must be involved and training must be continuously reinforced to keep skills sharp. This does not mean a single event training proposition. Continuous reinforcement is essential to keep the focus.

'Continuous improvements in any area eventually transform the operation. They lead to product innovation. They lead to service innovation. They lead to new processes. They lead to new businesses. Eventually continuous improvements lead to fundamental change' [35]. Companies like 3M are known for encouraging innovation. To remain competitive companies must develop such a culture. If your company doesn't the competition surely will.

> To be a successful change leader an enterprise has to have a policy of systematic innovation. And the main reason may not even be that change leaders need to innovate – though they do. The main reason is that a policy of systematic innovation produces the mindset for an organisation to be a change leader. It makes the entire organisation see change as an opportunity. [36]

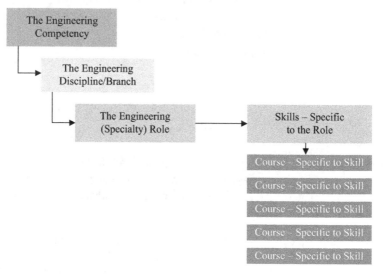

*Chart 3.16   Simple Development of the Body of Knowledge. Note: See Appendix
section for full page image*

3.7   Establishing effective administrative networks

Have we looked at all the networks that exist in the company? Have we looked at what works and what has been known to fail? Often failure is a nowhere place that we cannot visit. Which people have served as roadblocks and which have served as expeditors? Knowing who these people are, their roles, how they get things done and how they fail to get things done provides an organisation with a wealth of knowledge and ability that illustrates waste, value and improvement that can be put into the overall process. That knowledge and application to the process will facilitate the lifecycle of the product and the satisfaction of the customer. This analysis is also an issue of culture, because the actual culture of the organisation may foster their behaviour, be it good or bad. What is being done to change identify that behaviour, change it and improve the culture in which it functions?

Networks lend to the development of a new form of team, the IPDT. This is a concept that integrates the major players into a team based on their competency positions and expertise in the company. Ideally it coordinates the resources, communication, control, business methods and programme operations around a specific product or service. It differs from the traditional team in that it focuses and is organised around the product or service being developed rather than the disciplines to produce a product.

The IPDT's key features allow for a funding profile that permits early involvement of the appropriate disciplines, considers all product and process aspects, allows early participation of the stakeholders and has a hierarchy of product teams. Responsibility and authority are also delegated to the proper teams, who have ownership and accountability for cost, schedule and performance. Last but not least is the focus on customer satisfaction. IPDT has also been called concurrent engineering, with the benefits and a very similar process of organisation. Both the IPDT and concurrent engineering process provides the following benefits: time reduction for production of the product, cost reduction with less waste, lifecycle cost reduction due to longer life in place and the reduction of post-release design changes. IPDT and concurrent engineering allow for continuous improvement in the development of the product. This can best be illustrated by Chart 3.17, which shows how the continuous improvement environment supports all of the IPDT concepts.

3.8 The importance of developing a system for followership and membership

The culture and the networks all function together to show the planners where the followers are required or will be needed and how people will need to be encouraged to become members of the teams to complete the processes. Followers are those who can take orders, understand a process when instructed and carry out the instructions provided by the processes or the leader. As a team member, the individual has a responsibility to the total team process, understands the process parts and is willing to stand in when a member is called away. As a team, they work together to complete the tasks at hand and work towards the betterment of the process, and the product, again improving it and documenting the changes when approved by the CCB and management.

Followership is building a trusting relationship with your leader and the leader with the stakeholder or employee. A leader is worthless if no one believes or trusts in them. The leader must cultivate that relationship. They cannot take it for granted just because they have a position of authority, which only provides a position on the org chart. The power of the true leader–follower relationship comes from the followers, the real stakeholders, not the leaders. The leader cultivates that relationship by developing trust and 'walking the talk'. That does not mean that they have to be easy on the stakeholder and not demanding of the requirements. It means that the employees know what is expected of them and they are willing to follow this person because they believe what they have to say. They feel the relationship with their leader is predictable, consistent, ethical and based on values they agree with. Clearly, it is essential to communicate requests and expectations. Being realistic on demands also builds the necessary trust. This does not mean it's easy, but employees must have an acceptable expectation of what success is and how the leader looks for it. Hard work and long hours will be acceptable and even cherished if the relationship is based on trust and mutual commitment to common goals.

The same issues hold true for the team. As a team member, the individual has a responsibility to the total team, and understands the work process and its components. The collective team must develop trust in each member – trust that they will accomplish their assigned responsibilities in the time allocated and the manner chosen. If trust does not exist or has been reduced, members waste time and effort in making sure others get their work done, or correcting deficiencies by doing them for the others. Members must work together to ensure the work gets done by the assigned individuals. Each member depends on the other to get their task done. Their common goals are efficient and effective process output, and the betterment of the process. Ensuring future growth and documenting the changes made by team members is critical. This is why it is so important for the leader to be able to select their own team members based on the determined competencies identified from the WBS. As stated earlier, if each team member knows what their role is and knows what the others are to do and their roles, the trust factor gets stronger and the job gets that much easier to accomplish.

Based on this principle a similar study and project was conducted at the Northrop-Grumman Corporation. Don G. Freeman, Michael E. Hinkey and Jesse W. Martak, in their paper entitled 'Integrated Engineering Process Converting All Engineering Disciplines', stated that 'a benefit often overlooked is the impact on projects that are organised using Integrated Product Teams [IPTs]' [37]. IPTs cannot begin to achieve their true potential without an integrated engineering process. An IPT is essentially a team managing a small project. If each discipline in the IPT operates independently, then the team can hardly be considered as integrated. The team members are not functioning as a team if they are playing by different rules. The 'I' in IPT would stand for 'independent' rather than 'integrated'. According to Freeman, Hinkey and Martak, the organisation must look for processes that can be made common to all disciplines in order to promote process integration. In this system the CCB plays an important role. It establishes a formal release process for approving and updating the standard processes and all related processes. Does this sound familiar? It is exactly what has been said often in the earlier concerns for an integrated control system that allows an organisation to grow and maintain its purpose [37].

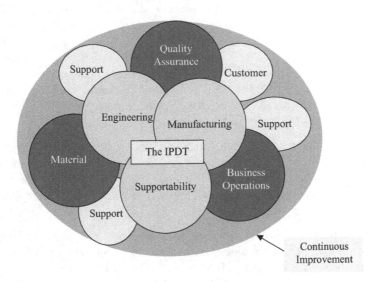

Chart 3.17   Continuous Improvement in the IPDT. Note: See Appendix section for full page image

This is particularly true in the knowledge industry, where no physical product may be produced. Relying and trusting that the team members will provide essential knowledge, facts or solutions when needed is the foundation of trust. Sharing information and helping each other resolve problems provides a strong foundation of trust. If an employee considers knowledge to be power and withholds it to increase their economic value, they will undermine the trust of the whole team.

The team is put together to get the day-to-day work done. Management gives it specific responsibilities and strategic goals to meet. Initially, the team is given budgets, equipment, people and so on. As the team becomes functional and self-managed it will negotiate with management for future budget and resources. The team selects its own leader who works with management to get the best from the team while negotiating with management for the work they are responsible for accomplishing.

Increasing team effectiveness is a double-edged sword. As they become more effective and productive, they become less tolerant of ineffective leadership. 'The talented people that surround a leader do not and perhaps cannot put up with a leader's (poor skills) for long. Abusive behaviour or inability to accept input is major impediments to team functionality. Peers and direct reports can easily undermine a leader's efforts if they feel that the leader's weaknesses are eclipsing his or her strengths' [10]. If a battle of wills develops between an effective team and an ineffective leader, the organisation will lose. Energy will be wasted by both parties. The team will redirect energy away from its primary purpose of productivity. And the ineffective leader will not help the organisation when they attempt to derail the team and micromanage it.

The more effective a team becomes, the less micro-managing they will tolerate. In contrast, the more control a manager attempts to apply the less effective the team will be. Managers who feel they are very effective, through their micro-managing and controlling, do not realise how little the employees are actually doing and how much more they could do if they were given the freedom to perform.

The relationship between the team and the outside leader must develop in its special way. As we have said before, the foundation must be built on trust. The team does not expect the leader to be perfect, just honest and open.

> Knowledge worker empowerment and freedom is a dual responsibility of the manager/leader and the worker. The freedom and capacity to make local decisions is vital to the IPDT and can only be assured through collaborative leaders working with their knowledge workers to prepare them for empowerment. [16]

Team members will work with leaders who work with them to improve. Honesty is critical from both sides. Both sides will feel safe enough working together and support each other. Honesty, respect and trust are foundations for developing the relationship between leader and team.

> One issue arises when managers give subordinates more freedom to make decisions without giving them the knowledge, boundaries, or context needed to make good decisions. Subordinates then proceed to make mistakes, feeling frustrated and even betrayed. The manager then withdraws the empowerment, convinced that it cannot work. The problem is that without knowing the limits of their decision space, without having been given the situational context and history of relationships, and without the experience of making decisions, no one can be successful. [16]

This is where the understanding of a person's capability is so important. By knowing this the leader can determine the need to coach, mentor or teach the stakeholder on those conditions that they lack. Again, when the leader develops their personnel because they know what is missing, trust blooms and the follower is ready to shine.

On the plus side, there is a multiplier effect when leaders and teams work together. In today's complex business environments, there are no simple solutions. Teamwork has already taken care of those. In most cases the answers to complex problems are not obvious. The leader or manager cannot be the hero and provide the answer. The team must work together with the manager, other teams and other resources to find the solutions. We're not talking about simple production issues or holding a kaizen event to remove or lower production costs. We mean problems that have no obvious answers. To solve today's problems requires all parties to trust each other and work together. If any party starts to pull back and protect their knowledge or position, the whole discovery process will fail [16].

To develop the effectiveness needed, team members and other teams must learn to work together. To accomplish this, organisations must change their informal reward system. Individuals need to be encouraged to contribute to the team rather than strive for individual recognition. In turn, the reward mechanism must not recognise individual achievements and successes but the successes achieved by the whole team. This may be frustrating for individuals who were taught to push hard to achieve individual success, but for the organisation as a whole it will have a long-term payback in overall productivity improvements.

The leader must have patience to allow the team to coalesce. They must decide how to work the problems and processes and who does what. This collective effort, if allowed to mature, will outperform any individual effort. Once the team has developed its own structure, it can take on large projects or small quick action efforts through assignment and delegation of the team members under any conditions.

'The successful team leader will walk a balance between self-organisation and leadership direction' [38]. The more control a leader takes, the weaker the team will be. In a complex environment, the leader wants a strong team, therefore it is to their advantage to allow or force the team to take control and develop its own strength. This does not mean 'free for all' management, but leadership by encouraging, watching and monitoring. The leader should only step in if the team is going to make a drastic strategic error. Allowing teams to learn from their mistakes will make them into stronger decision makers. A manager who looks for opportunities to wrestle control from the team when they feel it is in control will weaken the team and reduce (their) overall performance. 'Working with their people rather than over them, leaders can both liberate and challenge knowledge (which) works to stretch themselves, grow and contribute' [18].

3.9 The importance of teamwork in a process environment

When the unfinished product shows up at the team's area with the requirements set out in their processes, the team must utilise all of the capabilities they have been accounted for and apply them in the orderly approach required by the accepted processes and procedures. The procedures have been worked many times; hopefully all the wasted activities have been removed to allow them to function effectively.In a process environment, many people should know how the process is done and are capable of carrying it out.

As businesses move faster and work processes become more complex, leaders and managers must turn over daily operations of the productive process to those actually doing the work. Delegating to the team and providing them with the performance metrics necessary to monitor their own performance is a true performance enhancer. Resistance has been from managers who do not want to let go [39].

This shift in authority from upper and middle management to the work-force essentially means giving up authority while keeping responsibility – something few people are willing to do. Yet, to successfully release the worker's knowledge and experience for organisational improvement, the context, direction, and authority to make local decisions must be made available to all personnel. [39]

As in a military engagement, the closer to the battle line that critical, timely decisions can be made, the more effective and productive the results. The person in daily contact with the customer or vendor will also make the best decision. And just as in the military, upper management must give both the resources to meet rapidly changing demands. 'The probability of effective actions can be increased further if individuals work together to sense, interpret, understand and try to comprehend the environment. While the future may be unknowable, it is not unfathomable' [1].

When the raw material or unfinished components arrive at the team's area they must effectively process them and prepare them for the next position. They are trained to perform their operations by the leadership or management. Hopefully all the wasted activities in the process have been removed. In a process-thinking environment, people should know how the process is done and be capable of carrying it out in an appropriate manner.

Some managers may react with a 'sky is falling' attitude. On the other hand teams will be given more authority as they mature in the eyes of a true leader. As resources are always limited, funds will be managed by both the teams and the leader-managers. Negotiated agreements between team and leader as to how much authority they have will ease growing pains and settle concerns. There are never unlimited funds, but an agreement that addresses how it is managed reduces the friction.

This transfer of authority to the team will free up the leader and management to focus on efforts more appropriate for them, such as studying how to be more innovative, or what is the competition doing, or how can I make this process more effective? They can also work on metrics across teams and between the company and customers.

The TPS [Toyota Production System] working culture invests full faith and confidence in people doing direct work. It stimulates them to develop their capabilities to the fullest and makes maximum use of their talent. If leaders merely implement techniques without fully developing people, their system has no heart. [10]

Developing an efficient process is only half the job. Management and leaders must also develop their employees to perform efficiently and masterfully the individual duties required. When they are provided performance measures to evaluate their process output, they become self-sufficient. Leadership can focus on more valuable efforts and only monitor process metrics, rather than micro-management, which become a waste of time and effort for those at the upper levels of the company.

Empowering teams to effectively perform their processes and their assigned jobs is the real power of teamwork. The team members know their processes better than their bosses. They are in daily touch with the issues, problems and solutions of the process and their role. They can fix issues faster than their leaders or management. Build in a reward system which recognises team performance and effectiveness and the team will work harder to be more successful.

3.10   Wrap up on leadership in process organisations

Standardisation unlocks the power of consistency.

Standardisation ensures reliability and is the platform for improvements to take hold. ... Maintaining standards is an unrelenting job, and it is required of every manager to ensure that nothing is compromised. It is the essence of quality. It's what ISO 9000 and Six Sigma are all about. [2]

The best companies in the world are extremely orderly places, with little breakdown and disruption caused by failed systems. Leaders in those organisations spend time ensuring that policy and standards are maintained and if a failure or problem occurs, focused problem analysis takes place to identify the reasons for the failure to adhere to those standards. [2]

It is clear that what makes companies world-class. This clarity includes the following:

1.   An uncompromising attitude to quality and application to process.
2.   Agility, flexibility and speed to market of new products and services.
3.   Reliable resources (machines, people and systems), and
4.   The engagement of every employee in change and improvement. [2]

World class organisations have a plan that has a clear vision and implementation path. They understand how organisational alignment contributes to a quick response to an ever-changing environment. [2]

It is not uncommon for the Toyota employee to find they are empowered to act on an error in the process of operations and have the ability to shut down the entire line. This form of empowerment allows the stakeholder or employee to feel they are a part of the entire function and have a say in its operation. This engages the employee to encourage change and improvement. Agility and capability allow for more flexibility and there is no question that it has allowed new products to arrive at the market in a more approachable way.

'World class leaders understand that when a change over time on a piece of equipment is improved, there is a positive effect on their relationship with bankers and stock price analysts. This makes them passionate about the way folks on the shop floor carry out standardised tasks' [2]. It is also common for the same world

class leaders to recognise that reliable resources are a key to the company's productivity and gives them the very capability that they so passionately desire. Stock
prices rise and the value placed on the stock improves based on the analyst's
assessment of a more productive operation with a more capable workforce.

Questions for the reader

1. Does your organisation have anything similar to the 'hero'-type organisation referred to in this chapter?
2. What contribution does the Change Control Board (CCB) make to an organisation?
3. What are the effects of culture differences on an operating organisation? Can you identify these cultures in your company?
4. Does your company have any 'heroes' left in its ranks? How do they manage their current positions? Are they successful?
5. What changes have your leaders made in empowering the teams? Does your company have a team approach? How does it work?
6. What type of support is provided to those members of the company who work with associations to focus on new needs and sharing of ideas? Can you name any?
7. What are the five phases of the process engineering function? Does your company use any of them? Can they be installed in your organisation without too much trouble?
8. What does competency have to do with capability?
9. How would you use the competency model to develop your teams' capability and what would this do for your effectiveness?
10. What are the nine requirements for analysing the project, product or service? Does your company use these requirements? How are they used?
11. How do the contractual requirements differ from the analysis? Can you name the contractual requirements?
12. Why are non-contractual requirements often part of the agreement established by the company?
13. What are the five levels of the CMM? How do they reduce risk?
14. What do 'user-friendly' or 'ease of use' mean? Are these good terms to use in the development of a negotiated agreement to build something?
15. What role does culture play in your organisation? What are the tenets espoused in this chapter that encourage the leader to consider and develop an understanding of culture?
16. What is the role of followership in a company? Can you identify such a system in your organisation?
17. Is your company world-class? What criteria do you use to classify that condition?
18. Has your company developed its core competencies? What are they?

19. Are core competencies used to develop the employee's role requirements? Can you explain how this is done? Has anything in this chapter changed your mind about using and developing competencies and role requirements?
20. How would you define a successful team? What is necessary to develop such a team?

References

1. Bennet, A., Bennet, D. *Organizational Survival in the New World*. Amsterdam, NE: Elsevier; 2004, p. 185
2. Heymans, B. 'Leading the Lean Enterprise'. *Industrial Management*. Sept/Oct 2002, pp. 28–33
3. Bennet, A., Bennet, D. *Organizational Survival in the New World*. Amsterdam, NE: Elsevier; 2004, p. 76
4. Bennet, A., Bennet, D. *Organizational Survival in the New World*. Amsterdam, NE: Elsevier; 2004, p. 342
5. Bennet, A., Bennet, D. *Organizational Survival in the New World*. Amsterdam, NE: Elsevier; 2004, p. 77
6. Heymans, B. 'Leading the Lean Enterprise'. *Industrial Management*. September/October 2002, p. 29
7. Bennet, A., Bennet, D. *Organizational Survival in the New World*. Amsterdam, NE: Elsevier; 2004, p. 105
8. Bennet, A., Bennet, D. *Organizational Survival in the New World*. Amsterdam, NE: Elsevier; 2004, p. 297
9. Bennet, A., Bennet, D. *Organizational Survival in the New World*. Amsterdam, NE: Elsevier; 2004, p. 131
10. Cario, P., Dotlich, D., Rhinesmith, S. 'The Unnatural Leader'. *Training and Development*. March 2005, pp. 26–31
11. Bennet, A., Bennet, D. *Organizational Survival in the New World*. Amsterdam, NE: Elsevier; 2004, p. 132
12. Bennet, A., Bennet, D. *Organizational Survival in the New World*. Amsterdam, NE: Elsevier; 2004, p. 92
13. Bennet, A., Bennet, D. *Organizational Survival in the New World*. Amsterdam, NE: Elsevier; 2004, p. 183
14. Bennet, A., Bennet, D. *Organizational Survival in the New World*. Amsterdam, NE: Elsevier; 2004, p. 133
15. Bennet, A., Bennet, D. *Organizational Survival in the New World*. Amsterdam, NE: Elsevier; 2004, p. 134
16. Bennet, A., Bennet, D. *Organizational Survival in the New World*. Amsterdam, NE: Elsevier; 2004, p. 136
17. Kennedy, M.M. 'Career Strategies'. Presented at ASEE College Industry Education Conference, San Antonio, TX, ASEE, 2007

18. Bennet, A., Bennet, D. *Organizational Survival in the New World*. Amsterdam, NE: Elsevier; 2004, p. 139

19. Bennet, A., Bennet, D. *Organizational Survival in the New World*. Amsterdam, NE: Elsevier; 2004, p. 342

20. Csikszentmihalyi, M. *Finding Flow: The Psychology of Engagement with Everyday Life*. New York: Basic Books; 1998

21. Bennet, A., Bennet, D. *Organizational Survival in the New World*. Amsterdam, NE: Elsevier; 2004, p. 105

22. Drucker, P.F. *Management Challenges for the 21st Century*. New York: HarperCollins; 1999, p. 118

23. Software Productivity Consortium. *A Systematic Approach to Process Engineering*. Herndon, VA; 1999, pp. 2–1 to 2–25

24. Humphrey, W.S. *Characterizing the Software Process: A Maturity Framework*. Software Engineering Institute, June 1987, pp. 1–20

25. Bennet, A., Bennet, D. *Organizational Survival in the New World*. Amsterdam, NE: Elsevier; 2004, p. 21

26. Cario, P., Dotlich, D., Rhinesmith, S. 'The Unnatural Leader'. *Training and Development*. March 2005, p. 30

27. Robbins, S.P. *Organizational Behavior*. 9th edn. Englewood Cliffs, NJ: Prentice Hall; 2001, p. 515

28. Robbins, S.P. *Organizational Behavior*. 9th edn. Englewood Cliffs, NJ: Prentice Hall; 2001, p. 510

29. Robbins, S.P. *Organizational Behavior*. 9th edn. Englewood Cliffs, NJ: Prentice Hall; 2001, p. 512

30. Bennet, A., Bennet, D. *Organizational Survival in the New World*. Amsterdam, NE: Elsevier; 2004, p. 140

31. Robbins, S.P. *Organizational Behavior*. 9th edn. Englewood Cliffs, NJ: Prentice Hall; 2001, p. 512

32. Bennet, A., Bennet, D. *Organizational Survival in the New World*. Amsterdam, NE: Elsevier; 2004, p. 190

33. Landy, F.J. and Conte, J.M. *Work in the 21st Century*. 2nd edn. Blackwell Publishing; 2007, p. 593

34. Bennet, A., Bennet, D. *Organizational Survival in the New World*. Amsterdam, NE: Elsevier; 2004, p. 191

35. Drucker, P.F. *Management Challenges for the 21stCentury*. New York: HarperCollins; 1999, pp. 80–81

36. Drucker, P.F. *Management Challenges for the 21stCentury*. New York: HarperCollins; 1999, p. 85

37. Freeman, D.G., Hinkey, M.E., Martak, J.W. 'Integrated Engineering Process Covering All Engineering Disciplines'. Presented at SEI Conference; Pittsburgh, PA, 2002

38. Bennet, A., Bennet, D. *Organizational Survival in the New World*. Amsterdam, NE: Elsevier; 2004, p. 138

39. Bennet, A., Bennet, D. *Organizational Survival in the New World*. Amsterdam, NE: Elsevier; 2004, p. 11

40. Guderian, B. *Leadership Development and the Role of Continuing Education.* Presented to graduating class of the ELITE Program; Tulsa, OK, 2010, p. 2
41. Lareau, W. *American Samurai: A Warrior for the Coming Dark Ages of American Business.* New York: Warner Books; 1992
42. Frappaolo, C. 'Consultants View: Building a knowledge management program', *Beyond Computing*, 14 September 2000
43. Morrison, R., Ericsson, C. *Developing Effective Engineering Leadership.* London: Institution of Electrical Engineers; 2003, pp. 15–17
44. Bauer, E.E. *Boeing in Peace and War.* Washington, DC: TABA Publishing; 1991, pp. 151–63

Chapter 4

Maintaining product vigilance
and the need for change

Figure 4.1   Boeing P12 Bs and Cs (1934)

Maintaining an eye for product process improvement and the consistent awareness that we can change a process to improve that application applied to the lifecycle is the true measure of a mature organisation. Making sure that the changes are documented and supported is also a positive measure. Every organisation is aware of the need for change; it is the only thing we can be assured of that will continue to be a requirement in any company and for any individual. It is the common good that we often call 'growing' for any individual. However, making it a part of the overall company culture is another measure of maturity that cannot be ignored.

The Northrop-Grumman Corporation (NGC) made such a decision in their Electronics Systems and Sensors Sector (ESSS) in 2002 when their systems engineering, software engineering, RF and analogue engineering, digital processes, engineering support and other engineering organisations confronted the commonality of processes and sought to adapt their process requirements. That project included the adaptation of all the common processes, interrelated processes and discipline-unique processes that could be fostered as an integrated system. Freeman, Hinkey and Martak, key players in the NGC–ESSS team, had a vision that they would be able to merge five engineering components together in an integrated fashion that would facilitate an effective Integrated Product team (IPT) operation and would produce efficient projects [1]. These five components were: programme management, product support, systems engineering, software engineering and hardware engineering. Their near term vision was to develop and deploy a standard integrated engineering process, which included all the company disciplines and the supporting infrastructure with a process maturity that is the equivalent to level three of the Capability Maturity Model (CMM). Their long-term vision was to use quantitative process management to improve processes and increase the maturity of the organisation to the equivalent of CMM level four once level three had been achieved.

To accomplish their objectives and the vision that they had set, Freeman, Hinkey and Martak first established an Engineering Process Group, as suggested in previous chapters of this book. This was done with the full support of the NGC management and their encouragement plus support of all the other engineering organisations as well as those who worked with and supported the engineering functions. At the same time that they were looking at all the engineering processes, they were developing the necessary courseware to develop staff appreciation and user ability for training and education. They next established the Change Control Board (CCB) to monitor and establish the changes and approaches used to support the process functions. The first steps were (of course) to identify the key competencies of the company, then the major supporting areas (key process areas, or KPAs) areas and then the common and interrelated processes for each of the process areas [1] (Charts 4.1 through 4.7).

As they worked through the project, they made an interesting discovery. 'All the benefits that accrue to the software organisation apply equally well to the other engineering organisations' [1]. NGC–ESSS and the three authors discovered that IPTs could not achieve their true potential without using the real integrated engineering process. Individuals are not functioning as a team if they are playing by different rules. The critical tasks the authors discovered need to be emphasised:

1. obtain full senior management commitment,
2. establish an organisational infrastructure that supports KPAs,
3. identify KPAs,
4. identify common and interrelated processes in each KPA,
5. develop an integrated capability model (I-CM) that encompasses all KPAs,
6. develop procedures and work instructions for each KPA,
7. establish an engineering measurement charging system (EMCS),
8. define a procedure for project tailoring of the established standard process and
9. define an assessment process [1].

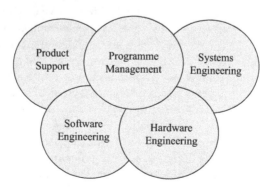

Chart 4.1   *Establishing the Vision at NGC [Source: Reprinted with the permission of Northrop Grumman Corporation. From Freeman, D.G., Hinkey, M.E., Martak, J.W. 'Integrated Engineering Process Covering All Engineering Disciplines'. Presented at SEI Conference; Pittsburgh, PA, 2002].*
Note: See Appendix section for full page image

The organisational infrastructure can only be established by leaders in senior management. This is why it is so important to have their total commitment at the start of such a project. The NGC organisation consisted of the engineering process group, the users group, a process CCB and pilot projects that reported to the Director of Engineering and Manufacturing.

The engineering process group at NGC–ESSS for this project was staffed full-time from each of the engineering departments or disciplines. It became the driving force behind the process improvement initiative. As shown in Chart 4.3, it developed the new processes, developed and inspired the training operations and courses, transitioned projects through to the new processes, maintained the organisation's process assets, and maintained the metrics, databases and training database.

The CCB plays the important role of establishing the formal release process for the review, approval and updating of the standard process for the organisation. Keep in mind that an appropriate capability model will not be available for each organisation that takes on this type of operation. Therefore, it is incumbent upon the CCB and the user groups to accept an existing model. Many choose the CMM 1.1 model to begin with and integrate the other models into its operation as an I-CM, as shown in Chart 4.4.

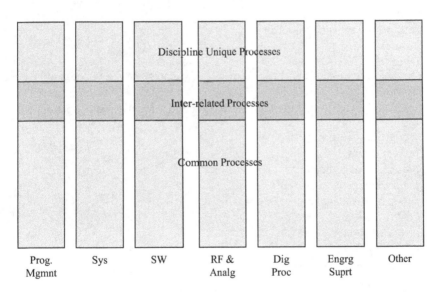

Chart 4.2 shows vertical bars labelled: Prog. Mgmnt, Sys, SW, RF & Analg, Dig Proc, Engrg Suprt, Other, with sections labelled Discipline Unique Processes, Inter-related Processes, and Common Processes.

Chart 4.2   Common and Interrelated Processes [Source: Reprinted with the permission of Northrop Grumman Corporation. From Freeman, D.G., Hinkey, M.E., Martak, J.W. 'Integrated Engineering Process Covering All Engineering Disciplines'. Presented at SEI Conference; Pittsburgh, PA, 2002]. Note: See Appendix section for full page image

The steps for establishing standard procedures and standard work instructions are established for each KPA. Each KPA is defined by three major products: the I-CM (Chart 4.4), which defines the process requirements; the standard procedures and the work instructions used by the developers [1].

Establishing the EMCS is an important element in the overall process improvement strategy according to the NGC–ESSS report. This views the EMCS as a record system of all the project labour charges on a daily basis. They record their labour charges using a standard predefined set of activities. This allows the developers to access the specific procedures or work instructions for each standard activity, therefore enforcing the use of the integrated engineering process. The NGC project personnel claim that without this EMCS it would have been impossible to compare measurement data between projects. The NGC EMCS captures the following measurement data:

1. activity in the standard 'activity list' that is performed,
2. lifecycle phase in which the standard activity is performed and
3. product being addressed [1].

NGC–ESSS and the trio of authors say that it is important to know how much effort is expended on the requirements, independent of when they are done in the development of the lifecycle. The present author agrees. The developer must also know or identify the type of effort as either:

1. original work effort,
2. work product inspection effort or
3. rework effort.

This identification allows the cost of work products and the effort to reduce the cost incurred to be determined.

Any EMCS should include the following standard activities as operational:

1. intergroup coordination,
2. project planning,
3. project management,
4. requirements management,
5. sub-contract management,
6. quality assurance,
7. project training,
8. process management,
9. quantitative management,
10. facilities and tools,
11. configuration management,
12. contract data management,
13. work product inspections,
14. product development,
15. product analysis,
16. product design,
17. product code and unit test,
18. all product testing,

19. product systems integration and test and
20. maintenance and operations [1].

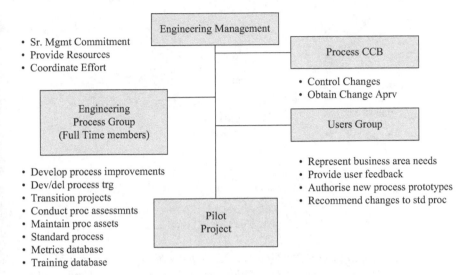

Chart 4.3   Engineering Process Organisation [Source: Reprinted with the permission of Northrop Grumman Corporation. From Freeman, D.G., Hinkey, M.E., Martak, J.W. 'Integrated Engineering Process Covering All Engineering Disciplines'. Presented at SEI Conference; Pittsburgh, PA, 2002]. Note: See Appendix section for full page image

Each of the key processes listed above are included in levels 2 and 3 of CMM 1.1.

Next, the CCB should establish the procedure for tailoring the project to the standard process. This process should become part of the command media, easily available to the process developers and requiring project managers to use the tailored version of the standard integrated process. Tailoring is not only recommended but required if one is to accomplish the development of each product in the life-cycle. By establishing the process the CCB enables each of the integrated development teams to identify the processes they will need in developing their individual product. They are then able to disregard processes that they do not need, allowing the team to function in the most effective manner.

Most organisations are structured in a stovepipe manner; this was also true of NGC–ESSS. This makes it incumbent upon the senior management to establish a comprehensive programme plan as an integrated engineering process that accommodates common interrelated and discipline-unique processes as discipline-oriented plans are replaced by activity-oriented plans. The comprehensive programme plan no longer has to exist as a physical document. It can now be accessed via the project's website and its contents accessed using hyperlinks and references back to the command media.

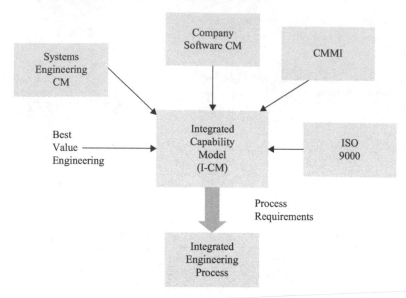

Chart 4.4   Developing the I-CM [Source: Reprinted with the permission of
Northrop Grumman Corporation. From Freeman, D.G., Hinkey, M.E.,
Martak, J.W. 'Integrated Engineering Process Covering All
Engineering Disciplines'. Presented at SEI Conference; Pittsburgh,
PA, 2002]. Note: See Appendix section for full page image

4.1   Learning to manage change as a way of life

Learning to manage change is not something that members of an organisation
automatically do or accommodate easily. We have to be taught how the change
process works and how it will work in a particular company. This requires a
management plan to establish training, budget for it and set aside time for
employees and management to participate and learn. Management has to take a
lead, and senior management must serve as key enforcer and information resource
for others. This is where 'walking the talk' plays an important role. If senior
management doesn't do as it says others should do, the event and its intended
effects will not take place or be enforced by others.

Management becomes the driver for the success or failure of change. As the
NGC–ESSS authors and operations discovered, without support from senior management there would be no change. And without that change there would be no
I-CM and the implementation that would follow 1. Again, it must be emphasised
that capability is the key to a successful company or organisation. If the capability
is a hodgepodge of skills that are not focused on the real needs of the organisation,
the success is questionable and will probably be a failure. Many organisations go
through the process of identifying the core processes and KPAs, but few follow
through with implementation and accessibility from management and personnel to
know where they fit into the grand scheme of things. Boeing has been one of the
companies that has successfully identified the capabilities of its personnel and has

made its skills inventory available to all of them. That inventory is there for stakeholder or employee evaluation and identification of the areas of development required to become fully proficient in their job or role.

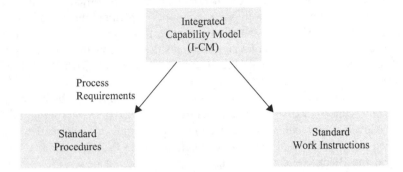

Chart 4.5 The Integrated Engineering Process [Source: Reprinted with the permission of Northrop Grumman Corporation. From Freeman, D.G., Hinkey, M.E., Martak, J.W. 'Integrated Engineering Process Covering All Engineering Disciplines'. Presented at SEI Conference; Pittsburgh, PA, 2002]. Note: See Appendix section for full page image

Change such as this is a monumental task and requires the full support of senior management. Without that support and influence the actual activity of identification of the core processes cannot be done. Next, the full body of knowledge that is required to support that core process must be completed. Again this requires the support of all involved – not just the senior management, but all employees who play a role in the production of that core item and the skills required to process it. This is where skills, abilities, knowledge, tools, processes, experience and so on are required and must be highlighted both for the good of the employee, by the employee and by the supervisor to assure that all is accounted for in developing that role.

Case Study: A process gone bad

Fertile Fields Corporation provides two major products to its buyers in four major retail outlets. Lately Fertile Fields has found that its customers have been returning more of their products for rework than they had in the past. Nearly 20% of the shipped product was being returned. The Vice President for Quality and Productivity visited Motorola while on a conference visit; he was impressed by their emphasis on their application of the 'Six Sigma' process and their ability to control defects in their products using this approach. Upon his return to Fertile Fields he made a presentation to the President on the concept and proposed that the system be applied to the products they produced.

Fertile Fields' plan entailed the establishment of a 'Black Belt' certification system to train and assign Six Sigma experts to the various production teams to implement the concept for product improvement. They would do this while operating within their usual quality assurance processes. A contract was let to Motorola to do the training and certification; members of the quality team who could be Black Belt candidates were identified. The trained and certified team members were then sent out to implement the overall steps into the quality process. The certified team members were instructed by the Vice President to look for faults and problems in the production process and work with the team to eradicate the errors that they found. Each Black Belt was assigned to a different production team.

The results were varied. Some production teams improved their quality levels and reduced errors. However, despite other teams also improving their quality levels and reducing the error rates, their products continued to be returned with definitive errors. As a result the Vice President simply discharged the Black Belts on those teams and constituted new ones. Nevertheless, the results continued to be bad.

Questions about this case:

1. What do you think is happening that makes a programme result in erratic production, when the very purpose of the process is to reduce erratic error?
2. Do you feel that the Black Belts were trained correctly? If not, what went wrong? Who should have done something about this problem?
3. Is there a problem with the Six Sigma process or is there a problem with the quality process established by Fertile Fields?
4. Following this, what would you do to improve the error rate?
5. What kind of problem did the Vice President create by removing the first group of Black Belts from the production process? Was his action appropriate? What should the Vice President have done, in your opinion?
6. Why do you think the production error rates continued to be bad?

This is change at its greatest position of importance. This is organisational change in the manner that it was intended to be established for the good of the company and the good of the stakeholder, employee and worker. This process is often forgotten or, more often, never attempted.

4.2   Developing an acceptance for change with adaptation and consideration

Change is accepted through training and education. It is a function that aids the attitude and ability that must be developed through mentoring, coaching and continuing education as a result of leadership involvement. Part of the education and training provided by the company must be to develop a positive attitude towards and about change and the process benefits that change brings to the organisation. The stakeholder must understand how it is done in the organisation in question. The leader can develop some of this through the coaching and mentoring process and especially by 'walking the talk'. If the leader does other than what they are saying, their stakeholders and followers will not believe or understand the message presented and will do other than what is desired.

Leadership plays a role in accepting change. The most effective leaders are capable of reframing the thinking of those they guide, coach, mentor and direct, so enabling them to see that significant changes are not only imperative, but inevitable. They can make stakeholders see the occasion as a chance to execute the leader's decisions and directives in a manner that has the greatest probability of success. This is again where the leader's example comes to the forefront and serves to develop acceptance by the worker and employee.

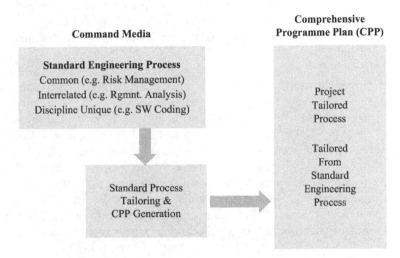

Chart 4.6   Command Media to Programme Plan [Source: Reprinted with the permission of Northrop Grumman Corporation. From Freeman, D.G., Hinkey, M.E., Martak, J.W. 'Integrated Engineering Process Covering All Engineering Disciplines'. Presented at SEI Conference; Pittsburgh, PA, 2002]. Note: See Appendix section for full page image

Executives and leaders who successfully implement change display the same basic behaviours, no matter where they are in the work environment. They often display basic emotions and a behavioural approach that fall into the standard patterns forming a structure of understanding and ability that describes the desired change process [2]. This is often the degree to which one demonstrates resilience or the capacity to absorb high levels of change while showing minimal dysfunctional behaviour. This can be seen by others as providing an implementation guideline that can be used by others during the change event. Much research has shown that there is a striking similarity in how people across the globe address transitions or change [2].

According to Conner [2], there is a set of eight patterns or principles used effectively by those who are responsible for influencing and carrying out key change events. These involve:

1. the nature of the change,
2. the process of change,
3. the roles played during the event,

4. resistance to the change,
5. commitment to the change,
6. how the change affects the culture,
7. synergism and
8. the nature of resilience of those involved (Chart 4.8).

Understanding how to use these patterns is essential if you are to successfully manage any transition or change event. It is not the events of change that confuse us as much as the unanticipated implications that these events bring to our lives. The crisis in life is the point where it becomes apparent that what we had planned to do is no longer feasible and our expectations have become disrupted or unattainable. This disruption becomes an ambiguity and the situation we find ourselves in seems to be out of hand [2].

The nature of the change concerns the momentum or complexity at which the change is presenting itself or the change is being implemented. With too much momentum or complexity, the receiver will become overwhelmed, causing dysfunctional responses. In a dysfunctional environment, each anticipated solution or attempt at a solution brings more complexity and dysfunction. These will require more creative approaches and a reduction in the erroneous behaviour. Just examining the simple fact of doing business in the environments where we try to deliver goods and services today provides the need for innovations in the information systems, organisational structures and production methods. These are causing challenges for change and the effective management of events. The shelf life of the existing solutions and operations are becoming much shorter, requiring new

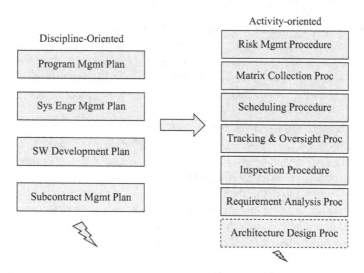

Chart 4.7   Replace Traditional Stovepipe Docs [Source: Reprinted with the permission of Northrop Grumman Corporation. From Freeman, D.G., Hinkey, M.E., Martak, J.W. 'Integrated Engineering Process Covering All Engineering Disciplines'. Presented at SEI Conference; Pittsburgh, PA, 2002]. Note: See Appendix section for full page image

problem-solving techniques and approaches to daily conditions from the leader. In *Future Shock,* Toffler warns that our effectiveness will be reduced if we don't learn how to manage and/or prevent the overwhelming events of momentum and complexity [12]. These are the change items that we face when we look at the hardware and software that we use to solve simple problems. Now, in our world of constant change, we are faced with technology that is changing at a rate of almost every six months. At one time it was a period of 18 months. iPhones, iPads and iPods are changing our world at an alarming rate. We have to accept the change, understand it, deal with it and put it to use in the best way possible, without allowing it to cause a disruption or other troublesome dysfunctional condition in our lives. But to ignore what is happening is not the answer. Understanding and applying resolve is the only real solution. This is what Conner calls **resilience** [2].

So, while Conner accepts that the nature of change is an important factor, resilience is the primary pattern that can be accepted. The other seven factors he calls issues or support patterns. Interestingly many of the other seven are not really supporting so much as they are affecting the positive or negative side of resilience. The support patterns, as defined, represent groupings of knowledge, behaviours, feelings and attitudes crucial to the outcomes of organisational change [2] (Chart 4.8).

These seven issues or support patterns make the stakeholder feel unstable or ill at ease in the face of change. It is the individual resilience that overcomes the issues and puts a calm face on the situation or event while it is taking place. Without resilience, the affected individual will become resistant to change and dysfunctional and will often bring up other defence mechanisms to resist what is being suggested or done.

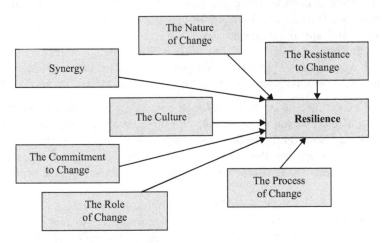

Chart 4.8   D.R. Conner Says That The Elements of Change are Best Met With Resilience [Source: From Conner, D.R. Managing at the Speed of Change. New York: Villard Books; 1994]. Note: See Appendix section for full page image

Conner states that the consequences of mismanaging change in one part of the world can directly affect many others around the globe. In a sense, quantum theory is here at its finest. Things will increase exponentially. Therefore, there must be some linking principles that either fortify or debilitate one's resilience to the effects of change.

With Conner's first element, the nature of change, we have to rely on the developed capability of the individual [2]. Again that similar component so often mentioned in chapters earlier comes to us: our skills or abilities, and our willingness to use them are always in play. These challenges provide us, in the form of momentum and complexity, with the opportunity to focus on whether this is a positive or negative event and whether this will have a disruptive affect on our wellbeing. The training one receives through mentoring, coaching and education plays a big part in understanding the event. If there is little understanding of what is being done, this will have a disruptive affect on the participant and will incite another pattern (to be discussed later). However, it is through the mentoring, education and coaching that we in the organisation provide which reduces the perceived disruption, changes the perceived expectation and reduces the feeling of loss of control that makes us become dysfunctional. Good mentoring, education and coaching programmes about the changes and what to expect provides an assimilation effect toward positive behaviour where stress and energy are diminished in a positive way.

Research has shown that companies that fail in their introduction to change have focused explicitly on hardware and software issues and not the mentoring, education and coaching. Successful organisations focus on the implications through mentoring of the new systems more than they do on the technology itself. So effectively we have to be aware of three forms in the nature of change. These are micro changes that affect you and those around you. They are considered to be organisational changes which occur at work and any other operation that affects your life. Macro changes affect you as part of the larger constituency. When change refers to 'I' it is micro, when it is 'we' it is organisational, and when it is 'everyone' it is macro, therefore the micro requires immediate priority and attention. [2]

At the same time, the intended goals of a company will not be accomplished or achieved if there are no micro implications for those involved. This puts the mantle of responsibility back on the leaders and senior management to realise they must establish a real scenario for the employee in which they will see the importance of the change and how it will affect them and their involvement with the product and the company [2].

Five key principles must be understood by those implementing change if they are to maximise the resilience factors of those working in the organisation. Number one is that we all seek control. If we lose it, disruption will cause fear and the avoidance characteristics we often see. Number two is that we all want some direct control over the situation. Number three is that we are able to assimilate the change at a speed commensurate with the events taking place. Number four is that we understand the micro implications and the macro effect. Number five is that we assimilate the demands within our absorption limits [2] – meaning that they can only be absorbed if we really understand what is happening and how it will affect us.

The process of change is Conner's second element. It has three phases, according to Kurt Lewin [3]: the present state, the transition state and the desired state. The present state is the current status quo. The transition state is when we begin to leave the present state. The pain of this condition is so great that we wish to move on to the desired state. When the cost of the existing state is so high it may become unacceptable, pushing us to want a new and better state. Eagerness to reduce the stress of the new transition state makes us more receptive to the new desired state even knowing how difficult, disruptive, time consuming and costly it may be.

Although it would be ideal for an organisation to see an opportunity to change and to take the steps to do so, it is a rare group that takes this process on without a lot of prodding. Organisations usually seek change and move into the realm of operational improvement due to two important business items: the high price of unresolved problems and the high cost of missed opportunities that we see in others' successes. Awareness of the IPT approach and CMMs and the use of process integrity might be seen as results of these losses and missed opportunities. These opportunities and the resulting changes needed must be stated in order to move into the new and successful business models. This was what encouraged both NGC and Boeing to move in the directions they did.

Boeing started early, probably more quickly than many other companies due to the forward thinking of their CEO at the time, Frank Strontz [4]. By 1986 they were up and running and organised out of the old stovepipe format into the new product and IPTs as a result of his leadership and support. In its guidebook *Total Quality Improvement: A Resource Guide to Management Involvement*, Boeing emphasised the importance of continuous improvement [5]. The essence of the plan is:

> To provide leadership in the continuous improvement process, managers must foster a climate of mutual trust and respect.... True quality improvement includes all employees working together in teams to search out the causes of various inefficiencies. This should be done on a daily basis so that quality improvement becomes an integral part of the way employees do their work. And it begins with the active involvement of every manager. [4]

While Boeing was not feeling any pain in terms of unresolved problems, Strontz felt it was missing out on lost opportunities. He proposed that his corporate team move into this new field of endeavour. It was not long after this that many other aircraft industries and those subcontractors serving the industry made the same drive to establish IPTs [4]. A major objective of the others, along with their drive for improved efficiency and operations, was to be compatible with Boeing. At this time the US Air Force was looking to improve its software processes. The Software Engineering Institute (SEI) was born at Carnegie Mellon University, with Watts Humphrey at its helm [6].

Everything we are talking about here, such as missed opportunities, is a matter of perception. Change will not occur unless there is a perception of need on behalf of those involved. It is incumbent upon the managers, leaders and senior management to understand where the employees at all levels are at the present time and are coming from in their current perceptions. The biggest failures will result when one approaches an understanding superficially and does something else other than what was agreed to by all the parties, encouraging the employee to recognise the falsehood of the situation and rebel. Successful leaders enhance their resilience with understandable processes that have phases that can be understood and anticipated. It is said that managing to the level of pain and the messages provided by the receivers is the first step in committing to change.

Conner's third component is the roles played during change. These are sponsors, agents, targets and advocates. The sponsors are those individuals who have the authority to sanction the activity. In the corporate community they would be the senior management. Agents are those individuals who are responsible for making the change or changes. The target is those individuals who must actually make the changes. Finally, the advocates are those who want to see the changes made but who lack the power actually to make it happen [2]. As in all real life situations, one must be careful how messages are passed between players. For example, it would be inappropriate for an agent to tell a target what to do. Sponsors cannot pass on their sanctioning power to people who do not have the status required by the targets. Therefore the executing staff (a process group) for the actual process cannot pressure line managers into doing their bidding without a top-down sponsor informing the managerial staff what it expects. This also allows senior staff to see what is being proposed and encouraged by the agents.

Case Study: The roles of change – Boeing vs. Boeing

Useem states: 'To less impassioned observers, it would appear that Boeing could use less "density" and more sense, after all, the days when technical marvels automatically produced marvelous profits are long gone; airline deregulation, the maturing of jet technology and – on the military side – the lack of a Soviet sized threat all mean that "higher, faster, further" has given way to "cheaper, cheaper, cheaper" as aviation's mantra' [14]. Some have even ventured to predict that with the acquisition of McDonnell Douglas (MD), many Boeing engineering and management types are feel that it was MD that actually acquired Boeing using Boeing's money. Some have called it a reverse takeover.

Peter Rhodes is a chief engineer at a Boeing subcontractor, Aviation Plus Inc. (both are fictitious and hypothetical for this case study). An important factor that has kept Peter working for this company has been the strong alliance of management with the engineering organisation. As the contract user, Boeing has always been an engineering-driven company, which has been to the liking of most manufacturing and engineering aficionados. The question that permeated the organisation in the past was 'What could we build to improve aviation?' With that the company and its subcontractors would set out to 'answer the mail' and provide customers with workable and desirable solutions. Peter is now in a dilemma. When his company presents ideas and solutions to management, before the customer is ever able to view the idea management is making adjustments according to a new type of category and question: 'Does it make sense to build this?' The question is usually couched in dollars and cents, and 'faster, better, cheaper' seem to be the battle cry. Cost-cutting has leapt over the importance of valued engineering. Peter now his engineering staff reluctant to take risks, to make valued suggestions or to provide anything other than what is asked for by management [14].

Questions about this case:

1. What is happening to the culture of this company? Does management see a need to support this infrastructure as the roles change?
2. What would you imagine is happening to the other support organisations?
3. Build a hypothetical scenario of the different departments in this case study. Establish the operating criteria for each department and explain how they would deal with the engineering and manufacturing departments.

Conner suggests that the 'sponsors always endorse the change project with the targets before you have the agents implement the change. And to the agents he suggests that they never take on a project where they are instructed to give orders to someone who does not report to them'. He further suggests that 'the advocates spend time with the sponsors, engaging in remedy selling and pain management' [2].

Sponsorship is more than a commitment; it is a dedication to the ability and willingness to apply the pressure and rewards that provide the desired results. If the sponsor believes that the change is a business imperative, they will be highly committed. If they understand how the change will affect the organisation in the short- and long-term basis, they are likely to sustain the commitment. The following five principles can enhance resilience:

1. understand and recognise the key roles in the change project,
2. be familiar with the effective operation of role configurations,
3. understand the general requirements of strong sponsorship,
4. recognise that change must be clearly sanctioned by all sponsor positions and
5. see that the rhetoric of change is consistent with meaningful consequences [2].

Resistance to change, Conner's fourth element, is a natural phenomenon. Momentum and complexity are compelling forces, as stated earlier. It is only natural that when confronted with the momentum of change, people cling to certainty and will oppose any interruption to what they consider to be working right now. It is now that we hear statements like, 'If it isn't broke don't fix it' and other clichés that ring true to the ear of the resister. It is a fact that when you attempt to establish a change, you lack the luxury of a single fixed reality. There will be many shifting images, interpretations and perspectives [2]. We can never forget that someone's perception is their reality. Having the ability to change therefore means knowing how to use the necessary skills and being willing to apply those skills as necessary. Remember that open resistance is healthy; an organisation cannot afford the luxury of covert resistance. Covert resistance is fostered by low trust, false statements and inadequate preparation for the event. Elisabeth Kubler-Ross said it best in *On Death and Dying* [7]: emotional highs and lows will result when people are forced to see the new situation following the loss of someone or something they love [7] (Chart 4.9). The status quo is one of those things.

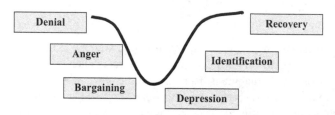

Chart 4.9   The Adjustment Curve. Note: See Appendix section for full page image [7]

At the very beginning of the project, employees must be told exactly what the true costs of it will be and why change is taking place. There will be surprises, but let the employees know that you are as surprised as they are. Senior management must evoke a sense of security by showing that they are also surprised by the discovery of events and in full support of what is being done. Being unprepared for surprises costs many individuals a lot of their precious time. So, anticipating them and working them into the process allows for a much more complete application, as well the understanding of the employees.

The fifth component is the commitment to change. This outcome requires an investment of time, money and energy. It requires participants to be consistent to the goal, even under stress. The process group has to resist ideas that promise short-term benefits that are inconsistent with the goals of the project. They must remain determined and focused on their goals, applying creativity, ingenuity and resour-cefulness in resolving problems and roadblocks [2]. Perception remains the most important commodity to commitment; the sponsor (senior management) must understand that they have to expend energy, time and resources to the effort. In this way the sponsor is using their organisational power to legitimise the change and ensure that it takes place. If the sponsor says they support it and then walk away from it, the project is doomed to fail. During the installation and adoption stages of the project the sponsor and agent must be aware of the logistic, economic and political problems that they can or cannot solve. Without resolution, the project cannot move on and will require the full involvement of the sponsor. Once the problems have been resolved, the institutionalisation stages begin and the processes for internalisation by the worker. This is done through the continued training and education of the workers and their supervisors over a period of time sufficient to resolve any resistance or concerns [2].

The sixth component is the culture of the organisation. This is the unique set of formal and informal ground rules for how we think and behave, and what we assume to be true about the company. In today's world of work, it is not uncommon for an influential person to work with an executive team that deals with expanding global competition, increasing customer demands, reduced financial resources, outmoded technology and customer service concerns. The major function of the culture is usually self-preservation. Its perpetuation protects the language, ideas, customs and manners of dress and behaviour unique to the society [2].

The corporate culture is a result of all the subcultures that have developed in the organisation in response to the unique challenges faced by different groups. Conner states that his 'research has shown the organisational culture to be the interrelationships of shared beliefs, behaviours and assumptions acquired over time by the members of the company' [2]. The following practices are said to be the true part of organisational life:

1. oral and written communications,
2. line and staff relationships, organisational structure,
3. how formal and informal power is cited and defined,
4. how time and quality are controlled and measured,
5. how formal policies and procedures are cited,
6. how reward systems and compensation are planned,
7. how the company's heroes, legends, myths, rituals and symbols are established and
8. how the physical facilities and their design are used [2].

It is therefore important that the change project is effectively aligned with the company's existing culture. It should be designed with a carefully constructed plan and beliefs, behaviours and assumptions that are consistent with the existing culture. The steps to developing a plan require that:

1. senior management and leaders define the specific characteristics of the desired culture,
2. all management and the leaders conduct a culture audit to identify any gaps between the existing culture and the desired one,
3. a detailed action plan is developed by the stakeholders to deal with the gaps and everyone is informed,
4. all management and the leaders then engage in a structured implementation of the plans encouraging the employees to participate and
5. everyone manages the organisational plan change in the culture and all the workers are informed [2].

The organisation's cultural traits must be consistent with what is necessary to drive the new changes and direction. Without this consideration, the change project will fail. Conner states that whenever a discrepancy exists between the current culture and the objectives of the change, the current culture always wins.

Conner's last element is the synergy pattern. Synergy is the company's influence on the other six patterns and the importance placed by the leaders on the final and successful outcome of the change project. It is the 'ties that bind' and either make them work together or at odds with one another (guaranteeing failure) [2]. There are two very important components to synergy: willingness and ability. These were mentioned earlier. Willingness comes only with the investment of hard work and perseverance that resolves healthy and productive conflict among the people with differing perspectives. Ability is the developed skills that are required to make the project work. It is often the teamwork required to develop the shared objectives, insights and ideas of the change project that forms the basis of willingness and ability. To be successful team members must demonstrate the skills required to be successful, using empowerment and participative capabilities following through to accomplish the requirements of the project [2].

As discussed earlier, resilience is the ability to absorb high levels of change while displaying the minimal dysfunctional behaviour regarding the change. Unconscious, incompetent people are unable to make the necessary adjustments to their lives when confronted with major change without displaying dysfunctional behaviour. According to Conner, 'most of the managers that he has encountered are unconsciously incompetent, he states that because they do not know how to guide their employees through the change process' [2]. This increases the importance of having an effective training and education component in the organisation that can deal with these unknowns, dig them out and present the discovered requirements in a clear and concise manner that develops the knowledge base. This is known as 'not knowing what you don't know'. The changes required to know the process alone are monumental. They require well-educated and knowledgeable personnel in the roles of sponsor, agent and advocate. Just one or a few knowledgeable persons cannot successfully establish and maintain a change project without a team effort. All the players must be knowledgeable [2].

Learning the human dynamics of change allows players to anticipate and prepare for the actions and activities that will take place. Most important are that everyone is aware of why this change is necessary and has a relatively good idea of what they can expect during the activity and after it has been completed. Anticipation depends on the ability to see how and when people will react, how they might express their resistance, how much commitment must be emphasised and attained, and how the organisation might influence the final outcome of the project. What we find is that resilient people react to change with the same fear and apprehension as everyone else; they just maintain their productivity and quality standards as well as their emotional stability while achieving their objectives.

Part of the human equation we have to deal with is that there are people who react differently to the environments and work situation that they encounter. These are Type 'D' personalities (Danger-Oriented) and Type 'O' personalities (Opportunity-Oriented). The Type 'D' personality views the crisis of change as a threat and feels victimised by it. Their tolerance for ambiguity is not well-developed and they view the change as an unnatural, unnecessary and unpleasant condition confronting them. 'They will use the tools of self defense such as denial, distortion and delusion to respond in a reactive non-proactive fashion', says Conner [2]. The Type 'O' personality has a strong vision, which serves as a template for the ambiguities. They will strive to stay on their plotted course and to maximise their personal resources for the common good [2]. They rebound well from the shock of change and will look for a way to adopt, adapt and move with the flow of change in such a way as to find meaningful approaches to what has been brought before them. They view change as a natural part of the human experience.

The key factor that the leader and an education and training organisation must build into the programmes that support the change project is the ability of the individual to understand and display a sense of security and self-assurance from the activities. The project must display a clear vision of what the organisation wants to achieve and why. The 'participants must develop a flexibility when responding to uncertainty with the ability to question in a positive way, searching for answers' [2]. The project must have a developed structured approach that manages the ambiguity, and most must understand it. Last but not least there must be an ability to engage the change rather than defend against it.

Managing the change is as important as the change itself. It is therefore suggested that the opportunity must be coordinated in the same way that engineers approach any condition: by identifying what the baseline is for resilience and resistance. Knowing this is not easy, but when we do have some idea, it allows us to identify which of the seven patterns will be the most useful in developing the desired outcome. These support systems should be established, and those which will be the biggest problems must be strongly identified, with plans to reduce them in the best manner possible. Having the processes and a true understanding of how to improve the productivity and meaningfulness of operations are so important that one must launch their improvement project as quickly as possible to fit the ELITE Leadership Model. Without the necessary approach to process improvement, a company will be doomed to inevitable failure in the long run. A new competitor will arrive in the near future and put you out of business while they use the new process and process improvement approaches.

4.3 Creating an entrepreneurial environment

If all employees felt they owned a piece of the company, especially the pieces that guide their processes, production and operations, they would do everything in their power to improve their productive steps. They would do more than work hard to remove the wasted time and effort in their actions and reduce the unnecessary procedures that they follow. This is why the entrepreneurial spirit is so important to a company.

Before leaving my role at Lockheed Martin Aeronautics Company, there was a great deal of discussion at various levels about what the new graduates lacked when they showed up at the door of the company to begin their new career. They had all of the pre-requisites, such as the mathematic background required to be a good engineer; their science abilities were also greatly appreciated and well-developed. But the biggest missing pieces were their understanding of leadership skills in an operational sense and a spirit of entrepreneurial applications towards new and undefined issues. In short, they lacked the ability to be innovative. Our challenge at the company was to build those capabilities and then drive them to the limits that we, as leaders and managers, desired. In most cases the leadership and management failed to understand what they needed to do, and often they felt it was not their responsibility to push that button. Training and education of leadership and management often fell short of solving the problem. This leadership misunderstanding required the issue to be brought to the very top of executive management. This is when the executive management insisted on a solution. Once this management level is on board, however, lower level management will take on responsibility to seek a solution as already described. It is interesting to talk with other management personnel in other companies such as IBM, NGC and Boeing. They also talk about the lack of these leadership and entrepreneurial capabilities in new graduates. All of the companies seemed to agree that they had to take on the responsibility to develop their new and incoming personnel.

Part of the entrepreneurial spirit is the ability to make employees feel that they are responsible for the success of the company – that they are major stakeholders, owners and developers in the process of things and can change and develop whatever is necessary to make the product better. This sense of ownership makes the company that much stronger and the employee that much more involved. Howkins [8] states that entrepreneurs in the creative industries need a specific set of traits, including the ability to prioritise ideas over data and to be endlessly in a learning mode. Developing this personal responsibility for leadership and innovation is desired in all companies.

In the course of events at any company, there is a desire to have the employee look out for the good of the organisation as though he or she owned it themselves. This form of responsibility is often built into the culture and presented as part of the company's history. With integral responsibility, developed leadership qualities and the ability to innovate within the desired conditions, an organisation has the best of all worlds. Acceptance of a CCB and the knowledge that they can bring requests to its attention, allows workers to present necessary changes to the processes and tasks at hand without fear of reprisal or ridicule. This makes for a good working environment.

4.4   Company history and the need for change

We have to know where we have been and where we have come from to understand why we do things the way we do, and often to understand why we are in this specific business at all. A lack of understanding about a company's history can only cause misunderstandings to escalate between old and new employees, as well as young and older workers. The importance of cultural awareness, leadership development for all employees and the understanding that change is an accepted condition are all part of an effective education and training programme, the leader's responsibility to mentor, coach and teach, and an environment that encourages change for the better as opposed to rebellion and rejection by the employee. Everyone must feel that a development process is taking place that encourages change or the ability to question existing criteria and processes.

Leadership is not just the ability to step up to the task and lead a group or organisation in the appropriate direction. It is the ability and resilience to deal with the elements of change that stem from an understanding of the operational processes and an understanding of the history of the organisation so that all motions made to integrate the appropriate processes will be embraced by those who live in the past, present and future.

There are all kinds of theories about leadership, including situational, functional, transformational and behavioural. However, the most applicable I have been able to find is the ELITE Leadership Model. This deals with the four parts of the present and future leader. These are (Chart 4.10):

1. A thorough understanding of the self. Do they have personal awareness, self-management, social awareness and the ability to manage relationships in an organisation.
2. People skills. Do they have the ability to develop effective teams, an understanding of how to develop people and motivate them, and the courage to deal with difficult people.
3. Understanding and ability to deal with the organisation. Do they have a handle on the company's vision, strategy and mission when working with the people that make up the organisation? An enterprising perspective that also understands the importance of change and the leadership required to manage that concept? And finally, are they aligned to the needs and requirements of the organisation based on its past, present and future?
4. Operational leadership perspectives. Does the individual understand the process imperatives involved in operational leadership? Do they have a grasp of the requirements of project management and systems engineering? What is their business acumen and are they applying wise judgment in the operations of the organisation?

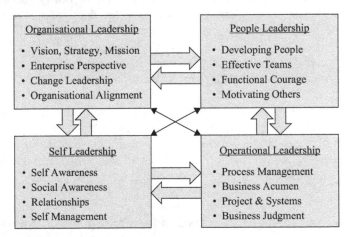

Chart 4.10   The ELITE Leadership Model [Source: Reprinted with permission of the University of Tulsa, ELITE Program [15]]. Note: See Appendix section for full page image

These four factors are probably the most important things one can learn in the process of understanding the true meaning of leadership [9].

With group dynamics a key factor in all of these components, one cannot overlook the effects of emotion on the leader and the participants as a process of social influence. It has been found that the more positive the leader the more positive is the performance of the group. It is therefore imperative in the course of events that the leader really understands the elements of process and their management. Employees will understand and follow positive and capable applications by a leader who knows what they are talking about.

Inherent in the elements of group dynamics is the question of style. It is quite obvious that a leader's behaviour will reflect a particular style that derives from their philosophy, personality and experience. Different situations also require a variety of styles. An emergency might require an autocratic approach to be effective. On the other hand, for a team that is highly motivated and well-aligned, a democratic or laissez-faire approach might be more fitting. The most damaging styles are those of a narcissistic or toxic leader as these abuse the leader–follower relationship. This negatively affects the performance of the organisation and the individuals involved [10].

This is where influence can work in favour of the leader. At one time we believed that a leader of a group had power and influence over that group. However, over time we have discovered that when one uses this form of power in a narcissistic or toxic manner, it generates the emergence of informal leaders who influence the group in other ways, such as subliminal resistance to the actions desired by the designated leader and organisation. The informal leader is not a friend to the group, but a reluctant leader who supplants the requests of the designated leader in a subversive manner. This is where the leader who knows the processes and capabilities can develop a following from each member of the group. Is this an emergency or a need for democratic styles of leadership? It cannot be forgotten that it is the leader who influences (this is the magic word – influences) a group of people towards a specific result [10]. The individual who is appointed has the right to command, but they must possess the adequate attributes and capability to match the authority with sufficient competence, recognise the ability of others and give them credit. This is reality, or the leader will be reduced to a simple figurehead by those who recognise the folly in their command or false leadership. Leadership is the ability to have others willingly follow. It is paramount, not just paramount for the individual, but also for the company to understand this phenomenon. Knowing who it is appropriate to assign to leadership roles is not just based on desire or wishful thinking. This activity must be carefully reviewed by the top management and implemented in such a way that the stakeholders understand who is in charge, how well they understand the requirements and how well the leader understands all the facts.

4.5 Lean management as a means for change and improvement

Lockheed Martin's fabled F-22 programme was an attempt to improve its production and operations by implementing the lean principles that it was promoting in its membership programme at the Massachusetts Institute of Technology (MIT). The idea was to use lean principles and approaches to produce a complex, novel product (the F-22 Raptor fighter jet) at low volume. This case study was published in March 2006 [11]. Lockheed Martin made significant investments in lean implementation, with the expectation of large savings. However, many problems emerged. The improved processes caused cost escalation, requiring leaders and managers to look at the process sources. Having taken corrective action and learnt its lessons, the company developed several insights and guidelines for a more effective process improvement environment [11].

The first problem arose with the new soft tooling concepts, which resulted in tolerance stack-ups that slowed the production line. The incredibly tight tolerances required on the F-22, and production's inability to baseline each assembly led to major problems for the production assembly line. As workers shimmed the parts that did not fit appropriately the tolerances began to stack up. However, for the F-22 to maintain its stealth configuration and remain aerodynamic it was imperative that the outer skin remain smooth and closer to specification. Based on its IPDT approach the designers were required to 'design for manufacturing and assembly', adequately accounting for the tolerances that Computer Aided Design would not. The lean implementation process was later credited with having caused these problems. The process was meant to have developed a rail system to maintain the tolerances, but it did just the opposite. It caused discrepancies in the assemblies' alignment, which resulted in a misalignment of the parts that were assembled at that station. The soft tooling that was implemented through the use of lasers failed to produce what the assembly operations required. The equipment took too long to set and often required two or more workers to do so. It was replaced with hard tooling, which was able to verify the accuracy, fitness and location of the problems. Also at the heart of the problem was a policy that had been written specifically to eliminate hard tooling and master gauges. To resolve this quickly, leaders came up with 'facility gauges', which were enough to solve the problems. These meant that shims were no longer required and the rail system became a transport mechanism. Soft tooling was re-evaluated as it became inadequate for consistent application to interchangeability and replaceability requirements [11].

The second problem was inadequate work instructions, which led to a lack of mistake proofing on the programme. The problem was discovered because of the lack of hard tooling, which meant that the requirements for quality were left to employees' experience. They spent most of their time looking at drawings trying to figure out what to do. It was clear after careful evaluation that unambiguous and more user-friendly work instruction would have saved a lot of time. The mistake proofing errors were discovered as a result of workers drilling most of the holes in the early manufactured aircraft by hand. Drilling through dissimilar materials burnt out drills and melted or splintered the composites, meaning many of the parts had to be scrapped. These problems were fixed by applying better training for the workers and establishing a statistical quality control programme to check the series of holes drilled [11].

The third problem was managerial in nature, as it required the leaders to do some hard thinking and planning. The lean efforts called for the elimination of a detailed manufacturing plan. Without the ability to look at production according to a plan, contingencies had no way to account for delays or tooling failures. In addition, management attempted to reduce the inventory to the number of parts called for in the bill of materials for each aircraft. No safety stock existed to cover mistakes, and in the beginning there were many mistakes, as already noted. As might be expected, part shortages created major problems for the schedule, and delays drove up the resulting costs [11].

Plans were put into place that expected the avionics hardware to be completed at least a year before it actually was finished. This meant that the software that was already completed could not be tested on the actual hardware. This led to a major configuration problem. According to the programme managers, this was one of the programme's major flaws. The problem leads to the lack of a production avionics lab for the flight line, due to the fact that a bad or non-functioning part had to be sent back to the supplier. This lead to the pirating of other working aircraft to further test the newest production unit and the lack of reliable equipment to ship finished aircraft to the client. The programme managers noted that it was difficult to continue to be lean in the downstream arena when this was causing a lot of problems for the flight line and its finished products [13].

The fourth problem was 'scope creep'. As the lean implementation was being put into place, new ideas and changes seemed to flourish from the stakeholders. While many of the suggested changes were helpful, they were not planned for and added cost to the overall operation. The largest of the unplanned changes was further lean and ergonomic enhancements. The production line received more electrical, compressed air and vacuum lines, as well as more raised access stands and application of the concept of the Six 'Ss'. The Six 'Ss' are sort, straighten, shine, standardise, safety and sustain, and come from the Japanese manufacturing processes used in Toyota plants. Although these concepts were important to improving the production, they increased the cost to the company [11].

The fifth problem encountered was the lack of learning by stakeholders from the problems faced in early production. So much emphasis placed on the production of the first 17 Raptors that the leaders and stakeholders failed to learn from their mistakes on those production articles. Those mistakes that have been mentioned were grossly overlooked, causing many problems for interchangeability and the installation of low-observable parts. The early, optimistic estimates of the learning curve and budgets were grossly overshot; actual performance increased the cost beyond what had been forecast. The planned savings did not come about [11].

As one looks back at the plans and the failures in the programme it can only be said that while it was the process that caused many of the problems, a failure to learn from the issues was the greatest downfall. Processes must be assessed throughout the function of a project, programme or service. As there issues that must be dealt with are discovered, they can provide a learning activity that should be shared with all the affected parties. This was not the case for the F-22. If workers had applied the process and discovered the errors, the correction should have been shared. There also seemed to be a lack of attention to what the stakeholders were telling their leaders. When the shimming problem started to have difficulties reach its true solution, the management and leadership should have worked with the stakeholders to resolve it, not to discourage them.

The ELITE Leadership Model states that process management, project and systems engineering, business acumen and good business judgment must be the rule for operational leadership. The leaders and management of the F-22 were not following these tenets, but then again they may not have known them. They were following concepts and ideas that had been put forward by MIT researchers and the industry partners of the Lean Management Institute at that University. These may not have included the importance of following the key processes as stated at level two of the CMM. It is more than likely that they had never heard of The ELITE Leadership Model. The problems described here seem to share a locked in assumption that nothing was wrong. This proved not to be the case, and problems probably did not raise their head until the costs went beyond what the leadership considered acceptable.

It is strongly encouraged that readers obtain a copy of the Lean Implementation study [11] and read it for ideas that they can use in starting up a new complex programme, project or service in their own companies.

Questions for the reader

1. What is the effect of change on the company when your stakeholders are not involved in the actual development of the need?
2. Why is the entrepreneurial spirit so important to a company? How do you feel about the emphasis of your company on the entrepreneurial spirit?
3. How does the leadership figure in the overall development of the vision, mission and goals of your company?
4. Looking at the Northrop Grumman Project, where major changes were made to the Capability Maturity Model and the resulting key processes, what would you need to do to get some of the same results at your company?
5. How entrenched are the organisational stovepipes in your company?
6. Would you consider it an emergency in your company to make the change toward capability or is there a need for democratic styles of leadership?
7. Does your company have a Change Control Board? If not, do you know why? If you do, how does it work, and is it working the way everyone expects?
8. Explain what an Independent Product Development Team is and how it works in the companies where it is used.
9. Can you explain the concept of 'resilience' and why it is so important to the establishment and control of change in an organisation?
10. What does entrepreneurial spirit have to do with the successful operation of a company? Do you see this spirit in your company?
11. What are the four functional leadership principles espoused by the ELITE Leadership Model? How are they applied to the function of an organisation?

12. What two elements have many companies identified as lacking in new recruits? Does your company also see this lack?
13. What is the difference between the 'Type D' or 'Type O' personality?
14. As you look at the case study presented on the Lockheed Martin F-22, give us your assessment of what you might have done if you had been assigned to lead that Program?
15. Of the five problems cited in the F-22 Program, which ones do you see as being normal in a new and just starting Program?
16. Does the concept of Lean Manufacturing appeal to you, even though it has been shown to be fraught with problems?
17. Can you name the lessons that you have learned from the Lockheed Martin study on Lean Implementation? How do they fit what you are trying to do at your company?
18. What concepts require a leader to be resilient in the face of change?
19. How much thought is put into a decision to assign someone as a leader in your organisation? Who makes that decision? Do they consider the capability of the individual?
20. A lot has been said in the various chapters of this book about the education and training responsibilities. How does your company deal with the introduction of change and is the education and training approach one of them?
21. How does your company deal with the criteria cited in this chapter about Mentoring, Teaching and Coaching? Does it believe in these concepts? Does it have a policy that encourages the leaders to do these things?

References

1. Freeman, D.G., Hinkey, M.E., Martak, J.W. 'Integrated Engineering Process Covering All Engineering Disciplines'. Presented at SEI Conference; Pittsburgh, PA, 2002
2. Conner, D.R. *Managing at the Speed of Change*. New York: Villard Books; 1994, pp. 84–5
3. Lewin, K. *Group Decision and Social Change*. New York: Holt, Rinehart & Winston; 1958
4. Boeing. *Total Quality Improvement: A Resource Guide to Management Involvement*. Boeing; 1986
5. Creech, B. *The Five Pillars of TQM*. London: Truman Tally Books; 1994
6. Humphrey, W.S. *Characterizing the Software Process: A Maturity Framework*. Software Engineering Institute, Carnegie Mellon University; June 1987, pp. 1–20
7. Kubler-Ross, E. *On Death and Dying*. New York: Collier Books; 1997
8. Howkins, J. *The Creative Economy: How People Make Money From Ideas*. London: Penguin Books; 2001, pp. 155–8
9. Guderian, B. *Leadership Development and the Role of Continuing Education*. Presented to graduating class of the ELITE Program; Tulsa, OK, 2010, p. 2

10. Ogbonnia, S.K.C. *Political Parties and Effective Leadership: A Contingency Approach*. PhD dissertation, Walden University; 2007, pp. 20–25

11. Browning, T.R., Heath, R.D. 'Lean Implementation Pitfalls in Low Volume Production of Complex Systems: Lessons from the F-22 Program'. Presented at Texas Christian University, M. J. Neeley School of Business, Fort Worth, TX, 2006

12. Toffler, A. *Future Shock*. New York: Random House; 1999

13. McManus, H.L., Millard, R.L. 'Value Stream Analysis and Mapping for Product Development'. *Proceedings of the International Council of the Aeronautical Sciences, 23rd ICAS Congress*; Toronto, Canada, 8–13 Sept 2002

14. Useem, J. 'Boeing vs. Boeing'. *Fortune Magazine*, 2 Oct 2000

15. Patricia Hall, *Elite Leadership Model*, University of Tulsa, OK, 2013

Chapter 5
The financial impact on process and operations

Figure 5.1   Lockheed mono plane visit to Schofield Army Base (1934)

Everything we do has a financial impact, we all know that. Or do we? The major thing we need to remember is that when we go through any change manoeuvre, it also costs something. We need to assess the cost that we exercise in order to save something on the other end, and the company needs to determine what that savings will be. Often we discover that the resulting savings far exceeds the cost of the change itself. In fact if we have done our business process re-engineering exercise appropriately, it will tell us what that estimated savings will reflect and the actual results will be very close to the estimate that we made. Over time the savings will often exceed our imagination and our estimates.

The financial impact comes with the cost of doing business and the need to make a profit. That profit will keep the company in the operational field where it can succeed. In order to be successful in business we have to know what things cost, that we have the most capable people doing the work and that the people know how to estimate and work with those costs in the best interest of the company and the customer. The customer really does understand that it costs something to build a product. They will pay what is fair for that product or service whether they are in the private sector or the federal sector. Bad processes and activities cost too much, as do less than capable employees who have to learn on the job or are trying just to get by. So when we start talking about key technologies, processes and capable employees we are looking to provide the best possible product in the shortest amount of time and with the best quality that these capable people can produce.

This is why we do all the things that we do. We determine what the best product or service is for the client by assessing the situation and then looking at the scope of that work (SOW) and the effort that it will take to build it. All of the tasks are then identified and the work breakdown structure (WBS) is constructed (Chart 5.1). The WBS is important because it helps develop the cost estimates to do the job and build the project [4]. Based on the requirements, deliverables, activities and processes that are going to be used, further knowledge about the real cost is developed. As the company develops the responsibility matrix for those who will be working the project, knowledge of their processes, working time from past performance and similarity of jobs done all add to the factors in play. This data also helps us do conduct risk analysis based on past experiences and draw up mitigation plans for the development of this very important endeavour.

What we are trying to consider is the best way to estimate the cost of the determined product or service. The best way to do this is to look at all the ideas that have been presented so far. Personnel hours for the best performers are the logical place to start. Keep in mind that some processes must be followed to do the most appropriate job. These are known as the operational processes – a key process in the ELITE Leadership Model. The time and effort expended is the cost based on the most capable personnel in use. This is estimated based on their cost per hour, the number of hours required and a forward escalation factor based on the lost opportunity because of the time in production, research and development, and so on. Forward escalation should be used with each resource consideration due to changes in salary, machinery wear and cost of repair (as well as down time during repair), changes to operations during production, and the loss of personnel (lost opportunity). Last but not least is the organisation's overhead rate. This is probably one of the highest cost elements in the mix. Cost of benefits and federal required costs are also all included, such as workman's compensation per employee, for example.

The next area is the materials used to develop the product. This comprises the actual cost of the material that will be used, the basis established on the cost of the use of the machines to build it and the escalation expected on the materials over time as it is purchased. This last is especially important if there is a long time element involved. It cannot be said too often that we have to recognise the escalating costs of materials and the repairs required on the machines as they age and wear. A burden factor also needs to be added, as the transportation, gas costs and personnel cost for moving these materials has to be considered.

What about the cost of random changes? This would include changes that might occur due to material changes, and processing changes as we learn to do things better with different resources. Without question, when there is a cost to building the product or service, that cost should be passed on to the client or consumer. Reductions in cost should also be considered for savings that might be realised when a process is improved and the cost of processing is reduced. Lot of groups should also be considered regarding the volume of the material used and the savings that might be realised from buying in volume. This was a change that occurred in the Lockheed Martin case study for Lean Implementation applied to the F-22 programme. When the volume changed due to errors in implementation the cost went up considerably, causing the programme to reconsider some of the ideas it was trying to put into place.

Two other areas are often left out of the mix when making cost estimates. Keep in mind that we are trying to follow the basics of the WBS and its stepwise manifestations for the totals to be considered. So far we have only worked in Steps 2 to 4. We have not even reached Steps 5 and 6. A real assessment of the schedule will show that we still have to add computer costs and personnel travel to the mix. What about those things such as the pieces or parts that we let out to our subcontractors? Their efforts and operations are going to cost us something as well. The burdens that have to be considered will be as elusive as those we have already looked at. Already the cost is at a questionable rate and we haven't even added the company's profit margin [4].

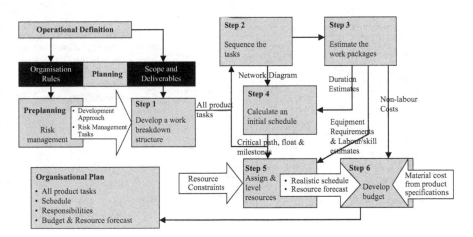

Chart 5.1   Developing the Work Breakdown Structure (WBS): Secision Making [Source: Reprinted with permission from Verzuh, E. The Fast Forward MBA. 2nd edn. New York: John Wiley and Sons; 2005]. Note: See Appendix section for full page image

With these numbers, we have what is called the 'top down estimate'. We now need to take a look at several other estimates. What do we think is our 'should cost estimate?' Do we have any idea what the customer is willing to pay or what their budget might be? We definitely have to look at that issue. Do we know what our competition's probable cost would be? Again, knowing this item is of the utmost importance as it will make or break our initial cost analysis. These are all good pieces of data for when the design team and company administration discuss 'go-or-no-go' decisions.

Chart 5.2 provides a suggestion for managers or leaders to use in transferring the estimated costs to the actual milestones of the product. From this point the leader can determine what they think the cost will be based on the major milestones and processes.

5.1   Being accountable for process finances

Every stakeholder should be responsible for the overall costs and savings that accrue from the changes and improvements of their functions in the operation. When a process is to be improved, such as when it is submitted to a change control group or board, the members of the team should be aware of the costs involved for this wait time. The desired changes and costs might be burdened on themselves, as stakeholder and requester. At the same time they should also reap the benefits of any savings that might result as a consequence. Everyone should be allowed to share the rewards whether they are positive or negative. This is accountability.

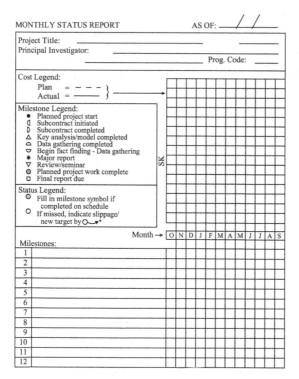

Chart 5.2   Monthly Status Report. Note: See Appendix section for full page image

As discussed earlier, the Integrated Product Development Team (IPDT or IPT) is necessary to accountability. Each individual representing the various disciplines must depend on each team member's ability as well as their skilled involvement in the effort. The teams will most likely be made up of individuals representing design, manufacturing, testing, software, systems engineering, quality assurance, programme planning and control, lifecycle cost and risk management, safety, reliability and maintainability. The working relationship within the team will depend upon the technical and interpersonal skills (their capability) that each person has. They will represent all levels of project development. Careful identification of the appropriate leadership cannot be left to chance; the leader's ability to develop a followership will give the team its determined focus and function.

The leadership of the IPDT and any sub-IPDTs will motivate the assigned teams only if they are chosen appropriately from the cadre of personnel available and represent their capability to lead. Capable leaders are often chosen on the basis of the following criteria:

1. company and product or service knowledge and experience with proven prior ability to lead multi-disciplinary teams,
2. company and product or service knowledge and experience with programme and business leadership; proven capability,
3. company and product or service knowledge and experience in the technical discipline associated with the team,
4. good working relationships with company management, the customer, personnel and other subcontracting company organisations and
5. company and product or service knowledge and experience successfully working with budgets, costing, scheduling and controlling operations.

While there will be many challenges to the leadership of each IPDT, it will be incumbent upon the selected leader to function as a facilitator and in many cases decision maker to ensure that rational processes are followed that are in keeping with company policies, while reconciling conflicting views within the team.

Accountability is determined by the 'prime' IPDT leader in order to ensure that the lower level teams and subcontractors are maintaining the cost and budget levels as well as the schedule and technical performance. This requires that the 'prime' IPDT leader conduct periodic programme status and control meetings. There they and the team will review the accomplishments and concerns expressed by the groups and encourage them to conduct self-evaluations of their operations and share them with the team members for improvements and corrections as the need arises.

In some cases benchmarking can and must be encouraged. (See Chart 5.3 for the benchmarking process.) Plotting other organisations' results against the performance of similar teams can be helpful. In addition, teams should be encouraged to set and monitor or track their goals through mutually agreed measurements. These must be supported for obvious reasons. Periodic meetings with management are also encouraged to discuss meetings or any inability to meet the goals set. This is also where the key process of peer review can be applied and used to increase success in the team's operations.

- Select the Process to Benchmark
- Determine the Project's Scope
- Choose Relevant Measurements
- Study Performance Boosting Best Practices
- Judge Appropriateness & Adopt Practices
- Identify Cultural Issues/Other Implementation Factors
- Plan & Implement Changes
- Measure Results & Analyse Benefits

Chart 5.3   The Benchmarking Process. Note: See Appendix section for full page image

None of this is possible without the most important ingredients of all: an effective training programme and a supportive top management that ensures that personnel will have the necessary skills and knowledge required to support the operation. This demands a concerted effort to ensure that key competencies are reviewed for every role and that any gaps are identified. These training and education programmes must focus on product delivery, design for lifecycle, facilitation, coordination and leadership skills, as well as the application of multi-disciplinary team skills to products, systems, sub-systems and sub-products. The training provided above all develops the skills to make the stakeholder accountable in their role for all phases within the lifecycle of the product, service or project. Again, the company will look to the leader to emphasise coaching, mentoring and teaching in addition to the company's own training and education programmes. Without this emphasis that supports this through "Walking the Talk," the employee will "not" see the requirements as necessary elements. In order to apply the appropriate processes it is encouraged that the leader looks to level three of the Capability Maturity Model and the key process of training. There they will find many successful suggestions and approaches that will result in a positive result for the company, the leader and the employee.

Arm-in-arm with benchmarking is the determination of best practices and the development of the business case, which focuses on why the activity is being done. Once the reader has determined who they will benchmark with – most commonly similar businesses – they will determine a best practice that they have been using to compare to those used in the other organisation. The attributes that they will be looking at will most probably be in the business focus, targets for comparison, rationale and metrics to be used, and what they consider to be the 'best in class' deficiency factors. These will determine the 'go-or-no-go' decision.

As the benchmarking progresses the leader and their staff will be looking at the cost of these best practices as well as their strategic or cost advantages. At this point they will be using the derived benefit or demonstrated improvement, and looking for a match with the business environment and the process applicability in the compared company or companies. The next step is to develop the business case on the basis of these findings. This involves determining the variable or value of transferring the best practice to the company in question.

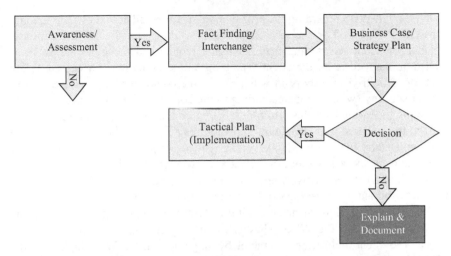

Chart 5.4   Benchmarking & Business Case. Note: See Appendix section for full page image

Determining the business case and transferring the selected best practices brings us to the natural steps we encouraged earlier. What are the basic requirements or resources required and what will be the strategic advantage of using this practice? Is there a business benefit? We now need to determine the net cost reduction or savings to the company and how these will affect the total recurring cost. The net savings in the first year and the risks incurred because of the changes will both be important to the management. Are the risks going to cost us anything?

Next we need to know how these changes will affect the programmes in progress and what our implementation targets will be. A business case to be presented to top management is then put into place. Who do we present the preliminary plan to and who do we need to have on our team as supporters and principles to support this change? Approval or rejection can be a major event for the company, so we need to be clear on the representation and support that we have and anticipate having. The presenters should be less concerned about any suggested revisions than about the potential of rejection. All potential problems must therefore be well thought-through and considered beforehand, based on what has been learnt from the benchmarking process.

Once the decision to go forward has been made, a target plan should be developed for implementation. This should include the approach to be used; the SOW; how the funding will be acquired, distributed and allocate; and the schedule. From here responsibilities should be determined and transmitted to the appropriate personnel.

It may be important to focus for a moment on those who work in the government contracting (GOCO) sector and provide a meaningful definition of operations for that arena. This is where the author's working operations mostly were. GOCO operations have a government client that requires the prime contractor ('the prime') to submit a full WBS regarding the project plan, detailing all activities. The funding to the GOCO is based on the best estimates of working man-hours, indirect costs (including subcontractors) and material usage required to complete the WBS as detailed and agreed to by the client. The programme execution is accepted by the client for a multi-year activity in which the prime contractor subjects themself to constant scrutiny (within limits according to the contract). It is also interesting that any savings identified by the prime during the course of the programme are, by contract, to be shared with the client. This requires prime and client management to review constantly for savings. However, in many cases prime management will cut budgets and manpower without the client's knowledge. This includes scenarios in which task and process analysis are not considered, much to the discontent of the manufacturing and engineering organisations within the prime.

IPDTs solidify client ownership with the GOCO by expressing support for the requirements through the WBS. They show that funding is based on agreements between the client and prime contractor by showing that the WBS highlights the tasks and activities to be adequately funded. However, in many cases what is often not shown and is not shared with the client is that the prime (as an organisation) has cut the funds based on unpublished savings, reductions in force and other profit motives that it has identified and are known only to the company. The manufacturing and engineering organisations are the usual groups to bear the cuts, causing them to feel that the original task and process analysis has been ignored and the new assessment (made by the company alone) is not being done and does not reflect the changes. This leaves manufacturing and engineering to do agreed-to work with the client with less funding and personnel based on the original WBS and assessment.

What has happened is that the budget and accounting organisations, as well as the company's upper management, have taken it upon themselves to make a powerful manoeuvre, instructing the manufacturing and engineering groups to continue their work with a cut to their available funds or positions. The manufacturing and engineering organisations are now required to cut tasks or processes indiscriminately to meet the new budget. Without meaningful reassessment of the tasks and processes, a true completion of the product cannot be done effectively or efficiently. Top-level management or a budget and accounting organisation should never be allowed to trump the requirements of the original agreement stated in the WBS. The client should be made aware of this change by the prime. However, without a desk audit of the operations, the client may never know. The prime should be expected to justify its decisions to the client, but does not, and instead expects the engineering or manufacturing organisations to continue as ordered.

Agreements signed by the prime and the client should include a requirement that the GOCO do a reassessment of the tasks, processes and activities whenever budgets or manpower are cut, processes are changed or activities are realigned by anyone in the prime organisation. The GOCO should be accountable for the changes and notification. This will help the client and the IPDT groups for the prime determine what is sufficient to the task, what will be required as the processes change and what activities will or will not take place. This will also support the most important detail that the client has purchased: the competency of the company to do the job as required. It is evident that when a client chooses a GOCO they do so mostly on the basis that the WBS will be supported by the most competent and capable personnel in the buyers' eyes. This is not uncommon as the client is buying the prime's capability and competency based on the WBS and its requirements.

The client, on the other hand, is implying that there must be a maintained capability by the prime, and that as technologies change a capability upgrade will be completed. The error is that many client project officers fail to conduct an evaluation that requires a baseline assessment of the GOCO's processes as the requested changes take place. This is no different to the company changing the budget without telling the client. If the client assumes that the required changes are being made, then there must be a way to verify this in addition to the subsequent desk audits that take place during the programme. If it is found that the company has not informed or does not have the capability and competent personnel to do the job, the client is obligated to request a plan from the GOCO to fix the deficit. Funding adjustments to a contract assume accurate personnel assessments. When personnel are cut, what is the sufficient task level and will the processes, as they have changed, account for the required capability level under the new criteria? Are there activities that will not be done because of the technology change or budget and personnel cuts? If the personnel assessments are not accurate, completion of the contract will be more than difficult. Problems incurred will include slowdowns, schedule change requirements and contract costs in excess of the original agreements. This is what most often happens. It is therefore incumbent upon the client to be accountable to increase the budget to accommodate the training and learning time for the staff as well as accommodate the waste that might occur during this period.

Profitability of the contract to both the client and the GOCO is determined by the accuracy of task, process and activity. This factor is determined by the accuracy of the personnel assessment, which must be equal to the requirements calculated as a result of the WBS. An inaccurate personnel assessment will create problems for manufacturing and engineering as they try to follow the contract requirements with lower than required budgets and fewer than required personnel. This also affects changes to the processes where technology has changed and the capability and competency of the personnel are lacking.

As shown in Chart 5.5, once the request for proposal has been accepted by both the company and the client, everything seems to start out correctly. To maintain this identity and impression of correctness it is incumbent upon both the company and the client to identify, continually track and help update the skills available to the working personnel. It should also be incumbent on the company to assess the critical skills required to complete the requirements of the contract, monitor the staff with those critical capabilities and ensure that their capabilities are kept up to date. This means monitoring the longevity, time with the company and the required promotions to keep them on board. The protégé list should also be scrutinised to ensure that critical skills are absorbed and ready when the most capable move to another job or from the company. This requires the maintenance of a 'watch list' of critical personnel and the continual upgrading of key staff to the latest core technologies.

The actions that support an efficient and effective budget alignment to personnel tasks, processes and activities are:

1. Core technologies required by the client should be maintained.
2. Core technology capability should be available as required.
3. Capable and competent personnel should always be available to the programme.
4. The roles identified in the WBS must be supported.
5. Personnel should demonstrate the 'body of knowledge' required of their role.
6. Necessary indirect support should never be cut below WBS recommendations or without approval of the client.
7. When improvements and cost savings occur, document them to support the changes, limit return to norms of the past.
8. Client and prime company should sign off on the changes and be party to the new WBS based on these changes.

Include an indirect budget and accounting person on the IPDTs so that the company can take ownership of the changes to ensure their adequate funding.

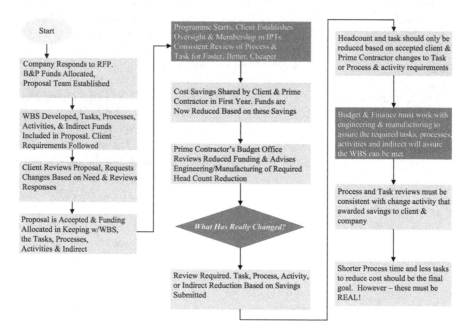

Chart 5.5   Assessing the Tests, Processes and Activities Needed to Complete a Contract. Note: See Appendix section for full page image

5.2 Direct versus indirect costing and accounting

We all work for years hearing the terms direct and indirect cost accounting. But many of us never really understand what they mean and often don't take the time to ask for fear it will show our ignorance. Since both those terms and their impact are important in the operation of a stakeholder organisation or team, everyone should be instructed and agree as to what they are and how they are applied.

Direct costs are those that are directly attributed to the design and development of the product or service. These include all the design activities and the manufacture of the product or service. Indirect costs are those which are considered to support the direct operating organisations as they build this product or service. These costs might include transportation costs, the cost of hiring the personnel and accounting or budgeting functions. It is interesting that it is often left up to the budgeting and accounting organisations to determine what is being spent on directly or indirectly manufacturing the product. There is no doubt that this arrangement is somewhat questionable since the budget and accounting groups are themselves major indirect organisations that support the company's operations. There are often squabbles between the engineering and manufacturing organisations and the budget and accounting operations as to what is an acceptable 'Direct' or 'Indirect' cost, and understandably so. However, when funding is completed and available to the company, it is incumbent that the company as a whole (engineering, manufacturing and support organisations) determine what are indirect overheads and what are direct applications towards the actual cost of building the project or service. Assessments of the indirect and direct activities, costs and budgets should be made as often as possible, but most importantly at the beginning of the programme when the funding is determined and received. The best assessment for determining the usefulness of any function or process used by a company is a value stream analysis of the operations. This can help determine which processes are used in both the indirect and direct operations over the product's lifecycle. Value stream analysis is also used to determine where waste occurs and how it might be avoided.

Value stream analysis is the best approach to meaningful applications and processes because it allows the organisation to evaluate the working activities in a process and to assess the value that it either adds or detracts from the product or service. Categories of waste to consider are:

1. overproducing – doing too much too soon, or quicker than necessary,
2. inventory excess – any form of batch processing,
3. excess wait time – too much downtime,
4. extra processing – re-entry and excess reporting,
5. correction or defects – errors and change orders,
6. excess movement – movement of material, equipment or people,
7. transportation – hand-offs or approvals and
8. underutilisation – ignoring abilities, authority and responsibility [1].

It suggests that value steam analysis is best done using the 'fishbone' approach to mapping out the flow of a specific process. The value-added tasks and activities are listed on one side of the fishbone and the non-value added tasks and activities on the other.

Chart 5.6 provides an example of a value stream analysis using the fishbone approach for the fabrication of an imaginary 'Furshluginer port valve'. The reader is encouraged to place the activities provided on a value added and non-value added fishbone diagram and determine if there are any activities that can be excluded. The idea of value stream analysis is to assess those activities that add very little value to the product and the process, and to evaluate whether one can do without the task altogether. Positive or value-added activities are generally placed on the top and non-value-added activities on the bottom of the fishbone. Keep in mind that all processes are made up of a series of tasks or activities. Each task that can constructively be eliminated realises a saving of time and money. In addition, where a task or a process can be improved the company and its client will realise a saving. The big concern is that if the company does not look at all the processes included, a larger picture of the overall process might be improved only a small amount or not at all, with the rest of the process unimproved [1]. This result does not give the company the kind of assessment that it needs. Value stream analysis should look at the whole process and not just a small piece of it. For this reason, the author wants to discourage the use of what many companies call, 'assessing the low hanging fruit'. This approach encourages looking only at a piece of the overall process and not the bigger picture. Be sure that when a leader undertakes a value stream analysis they are looking at the whole process and not just a piece of it.

5.3   Concern for stockholder return versus stakeholder investment and return

There seems to be too much emphasis on stockholder value these days, without a consideration for the real contributor, the stakeholder. Stockholders are looking for the success of the company to increase its value. However, this is only a fleeting involvement based on an income motivation for dividends; they invest in one moment and remove that investment the next. Stakeholders (employees, leaders, managers, subcontractors, etc.) have the greatest amount to win or lose in the operation of the company, yet we find a good majority of managers paying more attention to the stock and its performance than the means or systems by which we raise the return on the stock. That change in value comes through the more effective productivity of the stakeholder and the application of improved processes that contribute to the function of the development lifecycle and the longevity of the product or service.

Knowledge of the overall financial picture and the contractual means of getting things done help stakeholders do their work more effectively. For example, the more stakeholder personnel have a better knowledge of the contract procedures being used and the regulations by which they operate, the better they can do their job. This knowledge development can best be done through continuing education. Companies are always looking for information they can provide to stakeholders so that they can do a better job. Continuing education or coaching and development from the leader can accomplish this by informing stakeholders of the contracts and regulations they must follow. For example, what do they know about the contract administration process and what happens to the daily work reports? Do they have knowledge or understanding of how change orders are handled and why they are handled as they are? What is their understanding of civil rights and personnel issues, as well as the management of the materials that are processed in the company? How do specifications affect the company and fixed price versus cost contracts with and between the company and the subcontractors, and how can they manage the subcontractors better to ensure compliance to their agreements?

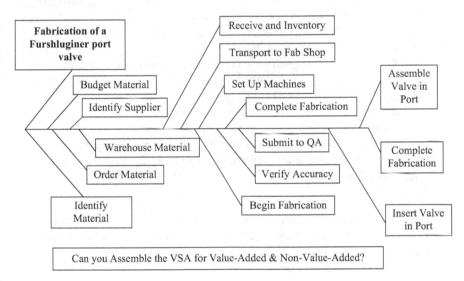

Chart 5.6   Value Stream Analysis. Note: See Appendix section for full page image

Case Study: Opinion poll [10]

Recently, a large manufacturing company in the southern USA wanted to conduct an employee opinion survey to verify its effectiveness in operations from the employees themselves. The Human Resources Department warned the company executives that they might not like the results. However, top management felt that it was doing a good job communicating with its employees, so went ahead to get the feedback. A section of the survey involved assessing the employee's trust of management. To the executives' surprise, the results showed that the employees trusted their immediate supervisors the most. The trust level declined the further up the management chain one went from the employee. Since this data was bad for the top level management, they refused to use the results. The data was not shared with the employees and was shelved. The results of the survey clearly showed that the memos and speeches by top management that had been intended to influence the employees had not had an effect or had it built what they wanted – trust. It was suggested to the management of this company that it took action, such as 'walking the talk' and making one-on-one daily contact to develop the level of trust expected [10].

Questions about this case:

1. Can the reader identify the major mistake that this company made by requesting its employees provide input and then not sharing the information and what they intend to do about it?
2. What kind of relationship between employees and company management do you think this event will foster?
3. In your personal opinion, what kind of action do you think they should take?
4. What would you do if you were a member of the top management level team?
5. What kind of action would you recommend to top management, considering the attitudes and opinions cited in this case?
6. Are the leaders or managers demonstrating the correct manner to handle the actions presented in this case study? Is there something else going on here? What might that be?

As well as contractual factors there are the funding sources. How much does the average stakeholder understand about overheads capital and customer funding? How much should they know, and have we even bothered to ask them? There is

Sales
- Direct Expenses

= Gross Profits
- Indirect Expenses

= Net Profits
- Tax & Dividends

= Retained Profits

Chart 5.7   Profit and Loss Statement Elements. Note: See Appendix section for full page image

also the factor of government funding for those who are dealing as GOCOs and the financial constraints that affect both entities. To be the most effective, stakeholders should be somewhat knowledgeable about the accounting processes and budgeting procedures used in each and every project or service. Personnel with this kind of knowledge provide more accurate and timely analysis of their work and perform better on the job. They are also more astute at allocating resources when necessary. Understanding the relationship between accounting and budgeting helps stakeholders to be more knowledgeable about the budgets and allocation of funds when they are necessary in the overall process. This knowledge has been shown to reduce the amount of waste that a company might accrue.

Continuing education of the stakeholder makes them aware of various means of budgeting, such as zero-based budgeting, bracket budgeting and activity based budgeting. Which one works best for their situation and helps them to do a better job? Understanding this approach through continuing education and knowledge development provides them and others with the ability to aid in the 'value stream analysis' by assigning cost and cash flow values to each of the tasks. Is the cost too much? Can the cost be saved? All this can now be taken into consideration and resolved while the process is in assessment. Using the form shown above in Chart 5.7, the stakeholder can now assess where the major costs occur and the effect they have on the company's cash flow. This knowledge also aids in the ability to stay current.

What a lot of stakeholders do not realise is that the business plan can be used as a leadership and management tool. The business plan helps the engineering leader track, monitor and evaluate the progress of the product or service based on the original plans established by the organisation. This assessment allows for alternative planning and realisation of possible obstacles that were not seen in the preliminary plan. Part of the continuing education process should be to help stakeholders see the means for righting wrongs that might have been generated during the original planning process.

For a long time leaders have depended upon profit and loss statements to see the results of their business operations. If taught correctly, the stakeholder can now see that the equation most often used in profit and loss assessment really means what it intends: profit = revenue − expenses.

Knowing these elements aids the stakeholder in understanding the economic value added (EVA) approach, which helps companies that are asset-intensive. The elements' primary advantage is in increasing leaders' focus on the activities that will increase the organisation's value. The EVA is calculated using the following formula:

$$EVA = NOPAT - WACC \times (Capital\ Deployed)$$

Where:

- NOPAT = net operating profit after tax (net income)
- WACC = weighted average of cost of capital (equity and debt) employed in producing the earnings
- Capital Deployed = total assets − current liabilities

Another way to look at the resources used in the operations of the programme, project or service is the use of what financial analysts call the 'Z Score' or the measure used to assess the likelihood of bankruptcy [9]. While the definition may sound scary, the tool is very good for telling the leader and their stakeholders how well they are doing in the overall scheme of things, especially with regard to performance. The original 'Z Score' calculations were determined by Edward I. Altman in the mid to late 1980s. His original calculations were quite extensive, but he later developed the four-variable 'Z Score' model [8]. The four variables are as follows:

1. 'X1', which represents the working capital over the assets times the coefficient 6.56,
2. 'X2', which represents the retained earnings over the total assets times the coefficient 3.26,
3. 'X3', which represents the earnings before interest and taxes (EBIT) over the total assets times the coefficient 6.72 and
4. 'X4', which represents the net worth over the total liabilities times the coefficient 1.05 [8].

To calculate a Z Score, first calculate the four ratios, each one being multiplied by their respective coefficients, then add the four values together:

$$('X1')6.56 + ('X2')3.26 + ('X3')6.72 + ('X4')1.05 = Z \text{ Score}$$

Once you have the score this can be compared to Altman's 'Cutoff Values' which are:

1. Safe, if greater than 2.60, or
2. Bankrupt, if less than 1.10 [8].

Kyd goes on to say that the Z Score takes a very stern look at your financial situation [9]. Therefore this author advises that the reader is careful about the use of this measure and probably should only use it if there is a fear of problems with the programme, project, or service operations. Using this measure will surely scare financial personnel if there is a result that shows the potential for an operation's bankruptcy. In the training exercise it should be pointed out that measuring the Z Score is to be used only in extreme cases.

For further definition, the following are used to ascertain the criteria used in the Z Score calculations:

1. Total Assets are defined as the available cash, receivables, inventory value and any prepaid expenses, including the net fixed assets.
2. Liabilities are defined as the accounts payable, notes payable, long term debt and other current liabilities.
3. Stockholders' Equity is defined as common stock value and retained earnings.
4. Measures used from the income statement include the sales, cost of goods sold (materials, direct labor, utilities, indirect labor and depreciation), gross profit,

operating expenses (selling expenses, G&A expenses), EBIT, interest expenses, earnings before taxes, taxes and net income.
5. Stock data that would be used would be the stock price, shares outstanding and market value of equity [8].

Now what do I do with this information? This is the most commonly asked question of those exposed to this sort of training. The best answer is to introduce the concept of the 'balanced scorecard' and to encourage its use. In its simplest definition, the balanced scorecard is a measurement or management system that is combined. It strives to create a model that can be used by the leader and the other stakeholders to bring in a very practical manner the financial data back to a strategic planning format with the activities used. The balanced scorecard recommends that the company's four perspectives are looked at:

1. financial,
2. internal business processes,
3. learning and growth and
4. customer.

Developing the metrics, collecting the data and analysing it relative to each perspective all have to be developed by the organisation doing the assessment [2]. This is because each organisation has its own specific characteristics. The assessment must look at these as specific to the organisation to understand where to go with the results and how to adjust for the best performance. Keep in mind that the balanced scorecard' is an assessment process that is used by the leader to analyse the ability of the organisation they lead. Each of the four items is subject to the analysis approaches developed by the leader, so whatever measures they wish to use to focus on any of the four perspectives are their own ideas. The concepts itemised below may be of assistance.

On the financial side, the company needs to succeed financially, answering the question of how should we appear to our shareholders. With this aspect we have to look at the objectives, targets, measures and initiatives. On the internal business processes side, the company needs to satisfy its shareholders and customers, answering the question of what business processes must we excel at. Again, the objectives, targets, measures and initiatives are examined. On the learning and growth side we have to ask what we need to achieve to meet our vision. How will we sustain our ability to change and improve what we do? And finally, on the customer side we need to ask the question: 'To achieve our vision, how should we appear to our customers?' [2].

Mentoring, coaching, teaching and/or continuing education should also look at the stakeholders' understanding of the raw material and inventory issues. For example, some training or education should take place that looks at just-in-time manufacturing processes and where they work best. Materials resource planning should also be investigated, along with the concept of manufacturing resource planning. Last but not least, the effectiveness or ineffectiveness of enterprise resource planning should be evaluated as it applies to the company. This assessment can be a real learning experience for the company – many learn that they don't need to follow the trends being led by the universities and training publications that taught these ideas. Often the assessment indicates that they are doing the right thing and avoiding the extra cost of following quasi-leaders from the outside world.

Process leadership cannot be effectively applied unless leaders or top-level managers have a good understanding of what these concepts can do for them – and if they are not used how they can avoid the pitfalls presented by the organisation, its personnel and the outdated methods used to produce the product or service. Production performance depends upon the leader doing their part in the planning, and communicating that plan to the workers. Some promote a seven-step planning process towards success [2]. This whole process begins with the establishment of performance goals and following through on those goals through feedback mechanisms and tracking procedures that the stakeholder uses. Evaluation systems are constructed and followed to assess the progress of the goals, and measures are made to verify success. This is also done with the aid of the stakeholder. As problems are recognised and corrected the system changes and new goals replace the old ones as the cycle of performance assessment continues. All of the pieces and parts that have been discussed earlier are integrated into the planning process and the development of goals. The performance management approach is illustrated in Chat 5.8.

Case Study: Image vs. visibility [10]

Peter Jones was going to a job interview in Detroit, Michigan with a national automotive parts distributor. He got his first glimpse of the culture when over the phone he was asked to drive to the head office location (a five-hour drive) at his own expense, where he would stay the night before the interview in a company-owned apartment. Peter thought 'OK, they appear to be frugal and that's good'. However, upon arriving for the morning interview he noticed immediately the image of the corporate headquarters as one with high flash and glitzy presentation. The facility was gorgeous and furnished with beautiful art and classic accessories. It appeared to be excessive for the lobby of a typical parts distributor's headquarters. Everyone there was dressed very professionally. All the women wore suits and all the men, similarly dressed, had monogrammed shirts, used Mont Blanc pens and had large gold watches on their wrists. It had all the markings of a marketing company.

When Peter went to lunch with the director he was interviewing with, he was told that he was expected to buy his own lunch – another inconsistency with the typical recruiting practices he had experienced, as well as the image portrayed. This was not a big deal to him, but the image of excess spending on the office flash, while behaving cheaply towards prospective employees was causing some concern. Peter was interviewing for a director's position, and as the interview progressed he discovered that the company wanted someone who had the marketing background to sell their product and a positive image of the company.

Since Peter had experience in these fields he further probed the director he was interviewing with about his part in the overall plan. The director stated 'Our philosophy here is "sink or swim", those that want to get ahead will find the means to develop their approach to sales. Those who do not develop and improve some younger, hungrier subordinate will happily replace them'. When Peter returned to the main office and spoke to others it was apparent that survival of the fittest was the management culture. Turnover in the ranks was expected, as well as that people would had to get ahead or be replaced. A win-lose competitive culture existed.

By the end of the interview, Peter's values, which emphasised developing management talent, and the company's win-lose competitive values were at odds and completely incompatible. Fortunately, both sides came to the same conclusion. The recruiting director sent a kind 'Thanks but no thanks' letter, which did not hurt Peter's feelings.

As Peter drove home, at his own expense of course, he could see how this car salesman temperament had created a culture that could only be tolerated by those of similar values.

Questions about this case:

1. After reading this case study, how compatible are your values with those of the automotive parts distributor?
2. How comfortable are you within your organisation? Can you re-examine your values and compare them to those of your company? How about those of your management?
3. Can you compare your values with those of your industry? How do they differ?
4. How would you classify Peter's handling of the situation? Would outward questioning of the issues done him any better?
5. What do you think the director's impression would have been when Peter told him that he was a development type of person, believing that anyone can be mentored, coached or educated to do what the company wanted?

Establishing performance goals requires the team to do several things. They must have a good knowledge of strategic planning and be able to look at all of the WBS components that fit in their operations. Next they must have a good understanding of the required tasks, processes and roles required to accomplish the goals they are setting. Have all the roles been identified and job descriptions set to assess the potentials for the roles? In other words do they have the best personnel on the job and in the roles that make a difference? What competencies will they be utilising and are these well described in the role descriptions? Has the company done a good job of identifying the core competencies and have they filled those roles with the best qualified? Last but not the least, is there an effective communication plan for discussing the goals, objectives and processes with the stakeholders, and what about the plan for discussing these same items with the new personnel as they come into the team?

The performance plan describes the desired results, how they will be measured and weighted and how they will be tracked by the leadership. Every aspect of the plan needs to be communicated to the stakeholders otherwise what is expected may never come to be. If you don't know what you are looking for you can be satisfied by anything; this is not what the company or leader wants [2].

In most cases the leader or manager is the one observing and looking for avenues of feedback from the stakeholders and workers. Information on the desired and undesired aspects of the plan and its operations are reviewed, recorded and noted for adjustment or replacement as necessary. Stakeholders are key to this function as they are the ones who must be convinced that something is being done to correct the problems and those who believe that certain issues should be discussed with others to guarantee that this will be done. Feedback or communication to those involved is a must.

The leadership must conduct the evaluation steps suggested in this approach on a regular basis. This is not a performance appraisal. This is the assessment of the roles that the stakeholders play in the plan and their projected results as a function of acting out the plan as presented to them. Often the measurement is done with a behavioural measurement tool that rates the behaviours and competencies being demonstrated by the players while performing specific tasks or

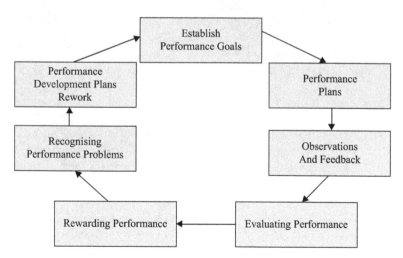

Chart 5.8   Performance Management Process [Source: American Society of Mechanical Engineers. A Guide to the Engineering Management Certification: Body of Knowledge. New York: American Society of Mechanical Engineers; 2006, p. 260]. Note: See Appendix section for full page image

processes. The tool usually provides the extent to which the performance meets the set standards [2].

Exceeding standards will often bring about recognition by the leadership of the stakeholders' performance. These observations should be noted and the team, group or employee rewarded by the leadership for their exemplary performance. This form of reward is a requirement. As stated earlier regarding rewards to the individual, the team and/or group will drive its performance higher and be more productive as a result.

When problems, such as shortfalls or deficiencies, are recognised in the plan, they should be noted and documented with recommendations from the stakeholder or leader to correct, or strategies for improvement by both. The goal in this step is to get the employees or stakeholders to suggest remedies to make the system and process better. This again will depend upon the results of the rewards provided by the leader, the company or the management.

The established performance development plan is the document in which to record all of the problems, correct accomplishments and new concepts to improve the process, operations or activities. With feedback collected at all stages, this plan sets the goals for improvement and regenerates the performance approach [2].

5.4   The negatives of top management salaries and bonuses

We are still rewarding the wrong people! It is not the top management or the CEO who succeeds in improved process application; that is, the reduced cost of production and the shorter time to completion. These individuals are a very small part of the team. If this is truly a team effort, then the functional team (those who do the work, the stakeholders) should be the function that is most highly rewarded. And it doesn't hurt to reward the most productive member of the team a little more than the others. That only stimulates others to improve their performance. The question that needs to be asked is: What has management done for the team that has made it so productive? In some cases nothing.

Throughout this book, the author has attempted to show that the team, IPT, or IPDT approach, together with the key and appropriate processes, will lead to a company's success. Yet the Boards of Directors of many of these organisations fail to see the animosity that is created when those people really being rewarded each year are the CEO and their immediate staff. There is no question that the CEO plays a part in the team and should be given a reward, but to make it so ostentatious only leaves the rest of the team members wondering what is happening and why they are so completely forgotten. The author also understands that it is the CEO who determines the product mix of the company and that they should be rewarded for the successful accomplishment of that operation. But again, it is not uncommon for a product to be cut from the organisation or sold to another company. And then it is lost to the fathering company altogether; those who work that product do not get rewarded or even acknowledged for what they have done in the past to make it attractive to the other organisation. However, what about the teams that do work the retained products and function well to provide what the CEO is requesting? As a part of the team the CEO should be sure to reward the members as well, for they have come through with a positive result based on the CEO's requests and leadership. Teams have to be rewarded, and the greatest contributors need to gain the best of the rewards. But to flaunt the salaries that are currently awarded by Boards to only the CEOs and their staff is not the way to encourage support from staff at the bottom of the organisation. It really rubs dirt in the eyes of those who have worked so hard. Is it any wonder that unions have grievances that hold up in court and often lead to personnel problems over the long term?

In a recent article published by the Associated Press, it was stated that 'the head of a typical public company made $9.6 million in 2011' [3]. This was an increase of 6% from the previous year. While companies trimmed their cash bonuses, they handed out increased stock awards. The article states that this is a victory of sorts; however, it is hard to believe that there is any sign of victory at all. It goes on to say, 'The CEO's motivation is to make sure the company does well'. It is important that the stock does well. However, this really depends upon the company doing well, and that can only be achieved if the stakeholders are doing what must be done as a team and making the product in the most acceptable way for the customer. To flaunt the size of the CEO's salary and other perks does not make the stakeholder happy, especially if the CEO is getting all the rewards but is not acting as a part of the team to accomplish the stated goals of the company [3].

Somehow Boards of Directors have to get a backbone. They must stop being 'yes' men and women for the CEO and look to the real good of the company. They need to make sure that the profits that they now give the CEO are shared with the overall team that made it possible – the stakeholders – therefore motivating the company as a whole to do better and to spend its extra cash on the things that have to be done. Those things should be to study the value stream and reduce the redundant and useless tasks now done in making the product. Too often the makeup of a Board is the result of the CEO's wishes and indications of who he or she wants to be there and how they want to be rewarded. That needs to change. There needs to be an emphasis on assessing the company's processes for improvement and making sure that the things that matter are improved at the source and conducted with the best personnel possible working based on their capability and experience [5].

No one seems to be doing anything about this obscenity. A lot of people are writing about it and raising the issue. In 2001 *Fortune Magazine* had two authors writing articles on what they considered to be an outright robbery: 'The Great CEO Heist' by Geoffrey Colvin and 'This stuff is wrong' by Carol J. Loomis. The point that the writers were making was their concern for the indiscriminate actions of those Boards of Directors that had a lack of concern and a willingness to allow pay levels to go ballistic [6]. One comment in the articles was whether we should feel sorry for those Directors who feel they have run head-on into a frightful dilemma. They seem to feel they have to pay big bucks to keep their CEOs or they will lose them to another company.

The Directors think they are doing one hell of a job. But they delude themselves. They think that things are being done right and fairly – they don't think that they are being had – when actually the excesses that they are approving are mind boggling. [5]

As members of the engineering and manufacturing side of the equation, we must consider the facts and be concerned that this continues. A CEO interviewed for the second *Fortune* article called the current situation a corrupt system [6]. His definition of the system was to the effect that it is 'Non-evil people doing evil things'. If that is how many CEOs or Directors are seeing it, why aren't we stopping it? In this author's opinion it is sucking the life out of many of the existing corporations, has done so for former companies such as Eastern Airlines and Pan American Airlines. We cannot forget the Enron debacle either. Funds that could be used to develop jobs and roles for others who are productive are currently being used in frivolous ways and for inexcusable means.

There is no benefit to this type of behaviour, which is often financed on the backs and hard work of dedicated stakeholders who are focused on the most effective and efficient process for turning out the product. Strong pride in a product can be destroyed when a few are perceived to be benefiting from the hard work of others and allowed to take advantage of the system. Perception is reality to those who see it in ways that can destroy the company. This is low value leadership at its finest.

As early as 1998 writers were expressing their concern for the obscene salaries that CEOs were receiving. In an article in *Air and Space Magazine* [7], Bruce D. Berkowitz points to what he believes to be inappropriate behaviour on behalf of the nation's CEOs. He points out that a managing director for aerospace research at Lehman Brothers saw that most aerospace companies at that time were reluctant to get involved in mergers. However, that reluctance was based on the fact that each company thought that they would be the major survivor of the mergers and therefore resisted consolidation. As new, more financially adroit CEOs took over, rewards were sought that were more financial than what many considered to be old world concepts of accomplishment. Being a financial hero was much more acceptable to the new CEOs than to those who had passed on. Of course, accomplishing this in the world of government finance was not easy. However, when they found that the ability to boost the price of their stock, discovery of that and selling off pieces of the company accumulated cash, and when they began to cut payroll they could funnel that money into the stock values and the new boosted stock value would allow them to raise their salaries. This also pointed them in the direction of buying other businesses, laying off their workers and pocketing the cash. They saw this as a means of reducing the number of companies and inspiring the stock market to look positively at their innovative actions.

It is interesting to note that it was Lehman Brothers who in 1998 coined the phrase 'for every Anders there is an Augustine', referring to William Anders of General Dynamics and Norman Augustine of Lockheed Martin [7]. Augustine was determined to be the survivor and dominate the aerospace market, where Anders was determined to get top dollar for his divestitures. Both were determined to come out on top. General Dynamic's stockholders were enriched by the proceeds of the F-16 production line sale. And this was only the beginning of the mania for merging and developing higher and higher price values on aerospace company stocks. Once the deals were done, the executives who had options to buy the company's stock at a preset price cashed in. Does this really demonstrate a belief in the company's ability or a money grab? Is there a benefit to the employees and stakeholders of the company that we are not seeing? Someone else is cashing in here and it is certainly not the stakeholder of the company who has put in the effort to make the company excel due to the quality of its product or service. Where are the real rewards to the real heroes of the company's success?

What we are seeing here is the law of the jungle. The uninformed participant should beware. Those who have worked to ensure the quality of the product and to do a job that was for the best of the organisation are on the outside looking in, while the manipulator of the system is accumulating the holdings and rewards. As one might expect the banks are not too far behind in support of what is happening. That is because when the cash flows it goes in two directions, and the banks are involved in both. Berkowitz went on to say that the equity analysts are also involved. These people are in the middle, advising both the buyers and the sellers on the merits of the consolidations and ensuring a piece of the pie to themselves as the transactions take place [7].

It is no accident that it was both the top management and financial managers who devised the process of performance management in the interest of promoting improved performance on behalf of the employee and stakeholder. It is time that we study this issue at each level of the organisation and look at who is really doing what for the company. Too much emphasis is placed on the lower levels and not enough on the upper levels – that might reduce some of the ripping-off that takes place. The idea of team management is seeing its time, and members of the team should each be assessed on who does what and how that effort will be rewarded. Maybe as part of the company review, it might be wise for the Board of Directors to look at team performance on the products or services it oversees? This would be a novel kind of assessment in which the various company teams actually get a chance to report to the Board on their progress over the past year, why they did what they did and how they helped their company improve and develop.

Questions for the reader

1. Name some of the costs that must be considered as the product and service is reviewed for an estimation of cost.
2. What does lost opportunity mean?
3. How is the burden rate determined?
4. What role does the customer have in determining the product or service cost?
5. What qualities are required for an effective leader in establishing the IPT?

6. Why is training so important to effective establishment of the IPT?
7. How is the benchmarking process used at your company? Does it work effectively?
8. Is there a difference between how your company's business cases are determined? Can you explain?
9. What are some of the differences that a 'GOCO' experiences that a 'For Profit' does not?
10. How does 'indirect' funding differ from 'direct' funding?
11. Can you determine the useful activities in the 'Furschluginer port valve' exercise and the wasteful activities listed?
12. Do you have a process at your company that would benefit from value stream analysis? What is it? Can you list the wasted activities that you might want to recommend to management as those to delete?
13. What does EVA stand for and how is it determined? How can this be of assistance to you as a leader?
14. How is the performance management approach used to improve the overall function of an organisation? Is there a good reason for using it?
15. Do you have a project to which you can assign the performance management approach, and can you do that project using these suggested requirements?
16. When you look at the executive compensation process, what is it that turns you on or off to its general application?
17. When should the 'Z Score' be used on a programme, project or service?
18. What does the 'Z Score' tell the user when the calculations are fully completed according to the formula?
19. How a 'balanced scorecard' is best applied? What are the integral parts and how do they differ from company to company?
20. Does your company have the best personnel on the job and in the roles that make a difference?
21. What competencies will they be utilising and are these well described in the role or job descriptions?
22. Is there an effective communication plan in your company for discussing goals, objectives and processes with stakeholders?
23. What about a plan for discussing these same items with new personnel as they come into the company and onto the teams?
24. Have you asked how you should appear to your customers to achieve your vision?'
25. Have the cost of random changes been considered in your programme, project or service? What are they and how are you planning to deal with them?

References

1. Keyte, B., Locher, D. *The Complete Lean Enterprise, Value Stream Mapping for Administrative and Office Practices*, New York: Productivity Press; 2004, pp. 6–8

2. American Society of Mechanical Engineers. *A Guide to the Engineering Management Certification: Body of Knowledge*. New York: American Society of Mechanical Engineers; 2006, p. 260

3. Associated Press, Rexrode, C., Condon, B. 'CEO Compensation Reaches New High'. *The Atlanta Journal Constitution*. 26 May 2012, pp. A12–A13

4. Verzuh, E. *The Fast Forward MBA*. 2nd edn. New York: John Wiley and Sons; 2005, p. 115

5. Colvin, G. 'The Great CEO pay heist'. *Fortune Magazine*. 25 June 2001, pp. 64–70

6. Loomis, C.J. 'This stuff is wrong'. *Fortune Magazine*. 25 June 2001, pp. 73–84

7. Berkowitz, B.D. 'The Wall Street Decade'. *Air and Space Magazine*. June/July 1998, pp. 43–7

8. Kyd, C.W. 'How are we Doing?' *Inc. Magazine*. Feb 1987, pp. 121–3

9. Altman, E.I. *Corporate Financial Distress*. New York: John Wiley and Sons; 1983

10. Morrison, R., Ericsson, C. *Developing Effective Engineering Leadership*. London: Institution of Electrical Engineers; 2003

11. Hall, Patricia, *The ELITE Leadership Model, The ELITE Leadership Program*, University of Tulsa, Tulsa OK, 2013

Chapter 6

How do we change – what do we need to do?

Figure 6.1   Pilgram-C24 at Schofield Army Base (1934)

The efforts that are necessary for the changes discussed in this book have been set out below for the reader to review. There is no question that something has to be done to introduce process leadership to your company. But what do you do? Several suggestions are introduced in a later chapter; however, listed here are some approaches that can be taken by the leader and the company that may improve the culture and environment.

- First, organisations must begin to focus on processes that consist of their key, core and important tasks or activities, and realise that these processes exist to produce a quality product or service. Companies should focus on the effectiveness of the production operation of these processes so that they will succeed as the organisation progresses through its required development. Moreover, the improved processes and their use must be viewed as essential to all their operations.

- Second, the standard rewards and punishments levied on stakeholders must be transformed or removed, away from old methods that restrict progress. Those rewards and punishments which have been in place since the inception of the company should be transformed. Or maybe the management likes operating in the industrial revolution? Today's more positive approach needs a new form of recognition of what is good and what is bad. We should no longer reward people (e.g. CEOs) for climbing the organisational ladder or chart as we have in the past. Neither should we reward people just for being very good engineers, manufacturing specialist or exceptional salesmen. There is a need to evaluate the goals the stakeholders have now set and the results they have achieved over the expected and measured period of time. In addition, who helped them accomplish this success? If we don't know who these people are, why not? Was it a team effort? How do we reward the team?
- Third, companies need to align rewards to the key, core and essential processes and goals of the individuals and organisation. People need to be rewarded for improving the productivity of the company and reducing variations in the production of the product or service.
- Fourth, the stakeholders need a fundamental reorientation toward a company culture that seeks to improve effectiveness and efficiency through productivity and service. This is required if the company is to advance itself and its staff.
- Fifth, the leaders of the organisation need to recognise that nothing will change unless they change themselves appropriately and develop a more positive attitude to coaching, mentoring and teaching the employee or stakeholder. Without this change of attitude the status quo will continue and the employee or stakeholder will not improve. To become more knowledgeable, stakeholders, employees or workers must have a more positive attitude about the company, the processes they are using and their leaders.
- Sixth, the company must identify their key or core competencies and make them an issue with which every stakeholder deals. The question must be asked as to who owns which competencies and how are the owners developing them? Is there a development plan in place to improve each one? What is the plan for identifying these new approaches and avenues that focus on those competencies, and are there related aspects that should be investigated? Are the stakeholders being kept up to date on changes to the key processes and tasks used to expedite the competencies?
- Seventh, the element of change must be accepted and adopted by all stakeholders in the company. Everything that is done must be looked at with a questioning eye and attitude as to how processes can operate better. What can be changed, and how does one go about making those changes? The role of the change control board (CCB) must be fully understood by everyone, and the CCB must be in place with appropriate policy, and must be used by the company as a tool to improve.

- Eighth, is accountability established for the financial elements? Who has responsibility for the finances and how will these they be held accountable? Accountability should not stop with the accounting and budgeting offices – they are really only there to support the engineering and manufacturing organisations and to make everyone honest. The key owners of the competencies and their financial roles must be known and understood, and the responsibility placed on their backs so that they hold the finances as a requirement in their review and control.
- Ninth, is the role of the company leaders in holding the subcontractors responsible for meeting the criteria set by the company in terms of requirements and processes? The leaders of the company must hold the subcontractors to the same level of quality and functional operation as they do the stakeholder or employee at the plant or job location. In reality the subcontractor is a stakeholder in the operation the minute that they take the contract offered to them by the project leader and the company. Leaders need to make sure that this subcontracting stakeholder understands the requirements and configuration set by the company and the expected quality for the production or service of their piece of the product.
- Tenth, tracking and oversight must become accepted and expected by both leaders, and when the production operations start to operate based on the production goals set, the leaders should be active in the constant evaluation of the progress of the products or services. The stakeholders should expect this and provide input where necessary to improve the process when suggestions are made by others. They should even make suggestions themselves for the good of the company. At the same time, the quality of the product will improve. This is where we find the use of peer review, as described in level three of the Capability Maturity Model Version 1.1 (CMM 1.1). Peer review is one item from the third stage of the CMM that the author encourages, as does the training key process. Peer review can also be applied even when the organisation is operating at level two of the CMM.

6.1 The first step

What are your company's core technologies? If you don't know them, then it is probable that there are a lot of employees who do not know them either. It might even be that the management and leadership of the company do not know them, although they think they do. This is really the first question that needs to be answered by any organisation. Without an understanding what its key technologies are, a company is unable to sell its capabilities to a client, or a product or service to its perceived customer. Based on the kind of work that the company does and the personnel that it has on hand, this knowledge will provide the first clues about some of the key technologies the company has at its disposal and deals with each and every day. It is this capability that the company sells to the client or customer. Capability is the skills, abilities, knowledge, processes, tools and experience of company stakeholders and those who can speak loud and clear with appropriate knowledge to the customer. If it were otherwise, those who succeed in making the company work would not be doing so. For example, why do you go to a McDonald's or a Wendy's Fast Food establishment? You certainly don't go there to get your car washed. Their reputation is based on their ability to provide the customer with the desired fast food, on demand and in the shortest period of time. When one of those establishments loses its ability to provide that key competency, its customers go elsewhere. It is no secret several sub-key technologies are employed and that the employees are trained for and expected to provide these abilities as a function of the key components. It is also quite evident that those who are well-trained provide the best service, but they also know what they are supposed to do and understand the customers' key technology needs.

In *The 21 Irrefutable Laws of Leadership* [1] John C. Maxwell points out several reasons why we are drawn to a particular product or producer. Some of Maxwell's are more important to the present context than others, especially regarding the first step. If we know what our key or core technologies are and we have hired stakeholders to support them, and we have a reputation in the field, then we meet many of Maxwell's laws. The important laws are those of influence, process, solid ground, magnetism, buy-in, momentum and connection. The law of influence refers to the factors of who people are, who they know, what they know, what they feel, where they have been and what they can do. You cannot do any of this without knowing your core technologies [1]. The law of process refers to important processes such as encouraging development of the self and others. It encourages the maturation of the self and others, since this is truly a culture issue. It encourages a change in the known abilities and capabilities of personnel, as well as a facing of the difficult issues of change in the culture. The law of solid ground refers to the necessary trust in the leadership, such as can you trust a leadership that doesn't care about core technologies? When credibility is questioned then the law of solid ground has been violated and is no longer in play [1].

The law of magnetism refers to the qualities that the leadership demonstrates by their actions. This says that the leader must generate the qualities desired by the followers or needed in a particular situation. These qualities emulate the attitude, background, values, energy and ability most desired by the followers, customers and clients. Buy-in and momentum go hand in hand. When followers, customers and clients have buy-in to the leadership, momentum starts to flow and move with the operational effort. Now the followers will do what needs to be done to get the product to market. Maxwell states that 'when you have momentum on your side, the future looks bright, obstacles look small, and troubles seem inconsequential' [1]. The law of connection refers to the ability to make connections that are powerful and relationships that allow others to follow without question. It requires the leader to make time to meet with his or her people, thus making themselves available to learn their names, to tell them how much they appreciate their efforts and most important to listen to their ideas, suggestions and concerns. Adding these types of connections to the reality of what one is doing will make most followers strong believers in what is being done.

Part of the awareness we have as leaders has to do with the understanding that there are four specific arenas in which we operate. One is the knowledge that we have built internally about ourselves and that we are capable of. The second is the awareness of our people skills and the ability to develop people to fill the needs of the company. The third is the lesser-known area of operational skills (focus on repeatability), the skills that we have attempted to develop in this book. The fourth is the organisational skills that the company may have developed since the stakeholder joined the organisation. This is the ELITE Leadership Model that we discussed earlier in the book. It requires a great deal of consideration if one is to succeed in the new corporate world.

The parts that we have spoken of most often have been the processes required to be fully operational and functional in the world of programme or project management. Those skills include knowledge of the requirements' management capability, the configuration skills, the programme management skills of planning, tracking and follow-through, and the skills of subcontractor management and quality assurance. These are the basics of the component we call operational processes, and systems management in operational leadership, just like the skills of project leadership, seasoned judgment and business acumen, are all required.

In addition, the ability to track and oversee the project needs strong emphasis. The skill and processes of planning in an efficient way allow for efficient and effective tracking and follow-through, thereby providing the leader with an effective set of guidelines to follow. Chart 6.1 gives details of the model. This principle is illustrated in the ELITE Leadership Model, developed for the ELITE Program at the University of Tulsa from an original model developed by Bryan Guderian of the Williams Company in his presentation to the graduates of the ELITE Program in the summer of 2011 [2].

6.2   The second step

Changing the system of rewards and punishment will not be simple. This is why the author has made it the second step. Without an appropriate rewards and punishment system the whole operation of the project, product or service is doomed to failure, right from the start. Too often we see organisations expect the impossible from their stakeholders, but when the final assessment takes place and the success is announced, rewards to the functionally involved personnel are hard to find. Apart from pats on the back and handshakes, nothing else comes along. But when the management incentive compensation packages are distributed to the upper levels, these accomplishments are lauded and compensated at often ridiculous levels. This is not a fair and balanced approach to supporting those stakeholders at lower levels who put so much work into making the product or project work for the company.

If management thinks for a minute that others, such as the employees or stakeholders, do not see and record this audacious act, they are mistaken. Those stakeholders who may have been in a leadership position, but not in the upper level management, view the acts as lacking consideration for the hard work they and others have put in to make the event or project successful. Lauding the top levels can be seen as audacious as the high rewards provided to the CEOs and their staff, who are most often seen by the stakeholders as those who may have had only a marginal responsibility and get the greatest recognition.

Chart 6.1   The ELITE Leadership Model [Source: Reprinted with permission of the University of Tulsa, ELITE Program]. Note: See Appendix section for full page image

The objectives, goals and team effort of the stakeholders most definitely need to be rewarded. To this end it is suggested that the organisation look very hard and carefully at its current reward systems and what they really want to support. As a result of this evaluation it should establish a new set of standards that can be measured and compensated to the team, its members and the leaders of the effort, project or service event. The new rewards system should also look at how standard punishment is handed out to those who fail to do their job or role and as a result are not successful in their projects or in similar events.

Motivation for success is based on what team members and stakeholders see as rewards or punishment. If the rewards are there, but are withheld because of a lack of success, then there will be a general understanding of the 'whys' and the leaders will have justification. Leaders are expected to demonstrate their capability to accept failure as well as success. Therein lies the real leader's ability to look at what was expected, and to see what was not achieved. The leader can and must explain the results in the most realistic of 'terms', relay the unhappy results to the team and still hold their loyalty over time based on the attitude of positive thinking that they have developed and the realisation that the team will do better next time and really mean it.

However, when a member or some members of a team are randomly laid off after the unsuccessful completion of a project, this only broadcasts the wrong message to the rest of the team. While they will know that the expectation was greater than what was delivered, the team members or stakeholders will see the contributions of those who have been reduced in force and ask 'How do I do my job, be less visible, but become less vulnerable to lay-off?' That's not a question you want stakeholders to be asking during or at the end of a product, project or service. This is the reason for establishing the rewards up front and withholding them only when the finish is less than acceptable. If layoff is a means of punishment for non-production or non-service, this should be clear at the beginning of the event, not a surprise at the end of it. Then the punishment is understandable, unquestionable and accepted by all concerned.

6.3   The third step

Are we rewarding stakeholders for the right things? We set the goals and forge the mission, but are we rewarding them for staying on target and moving in the direction that we set for them? Many organisations identify their key, core or driving processes and produce a list of them for all to see. However, do they make it clear to the stakeholders that the expressed importance is to operate at a specific level and to improve the processes as necessary over time and operate with the necessary vigour to save the company money, operating costs or expenses as they work through their projects, products or service?

If there is no reward for doing what the company expects, no operations will hold the key, core or required functions as important, and the processes will be completed only as the stakeholder sees fit based on their personal determination of importance. This is an assumption, but sometimes assumptions are based on perceptions and not reality. Assuming concepts that are understood is not a perception one wishes to leave with the stakeholder. Even when a perception is shared, it is not enough to simply tell someone what it is; the leader must 'walk the talk' and explain again and again to reassure that the stakeholder understands what must be done. When the company starts to reward its stakeholders for doing the right thing, then it will see the processes being used and the core technologies come into play.

It is incumbent upon the leader or manager to know what the company's key processes and technologies are. With this knowledge, they will be expected to mentor, teach and coach stakeholders, as well as 'walk the talk' in their use, and to emphasise the value of observation for each variation the stakeholder can find to improve the processes over time. The leader should be utilising the methods available to reward the successful and to provide incentives at every step to motivate stakeholders. Focus on the processes and their application will improve, as well as the ability of the company to apply its best personnel to getting the job done in the most efficient manner.

In a study conducted by Sibson Consulting [3], Conlon, Isler and Kochanski found that employees returned more valued contributions to the company if their rewards followed the employee value proposition (EVP). This represents and measures the rewards employees feel they receive [3]. An interesting result was that the stakeholders felt that work content, affiliation and career development outweighed benefits and compensation. As rewards go, this turns the approach that many companies use to reward and keep their employees on its head. Most companies believe just the opposite. Where work content, affiliation and development were emphasised, the stakeholders' level of engagement increased. This in turn had a significant impact on the business outcomes of the company. On the work content element of the EVP, the most important items to the employee were the skill level, variety and capability of the leader who was holding the stakeholder accountable, and job responsibility as determined between the leader and the employee. This strongly emphasises the concern expressed in this book for leaders to know the key and core technologies and processes and understand the body of knowledge for each role when assigning them to stakeholders.

On the affiliation element of the EVP, the important points to the stakeholder were the organisation's reputation, an understanding by the leaders of the vision, and the commitment and support provided [3]. Trust in the leadership was also strongly chosen by the participants of the study. So when a company starts to cut corners and its reputation in the field starts to slip, should the leader take notice that the stakeholder might also be watching? A company's reputation is of the utmost importance to the stakeholder, so to maintain trust, the leader must maintain the ability to portray a positive business image to the field.

On the career element of the EVP, the employee's title satisfaction rated the highest, with job security a close second and training and education level third. Where the leader has provided a job or role title incorrectly, the satisfaction will be less, and where the potential for loss of a job is greatest, the satisfaction will also be less. Again the rewards component of this step is important; we must make sure that the stakeholder in the specific role is capable of doing the job and feels well titled in that role. Too much emphasis on termination will most likely have an effect on the job being done. The Sibson Study found that just over one half of those involved in the study felt engaged [3]. This meant that they knew what to do, they understood the vision of the company and demonstrated a commitment to the company. The study went on to point out that where almost half (44%) of employees have low commitments to the company, performance falls short of what most employers' desire. It must be emphasised again that rewards for the most positive activity must be part of the company's actions [3]. The author states that the reward must reflect the fact that job content, affiliation and career are of the most importance, leading to the need for rewards geared in this direction.

6.4   The fourth step

This step requires the use of a thorough analysis and application of a re-engineering process to the culture. For culture is a process established over time and drilled into the personnel as the way to get things done. In a company, it can often become a roadblock. The importance of establishing a culture set on improvement is of the utmost importance. Improving the productivity and effectiveness of the organisation requires a whole new look at how the current culture was established and how one will re-establish a new one.

In *Re-Engineering the Corporation*, Hammer and Champy [4] state that when a process, as with the culture, is re-engineered, jobs evolve from narrow and task-oriented to multi-dimensional. Work units change from functional departments to process teams, stakeholders' jobs and roles change from a controlled situation to one in which they are empowered and understand the role they play. Job preparation changes from training for specifics to education on the broader scale. Performance measures and compensation shift from activity to results. Advancement changes from performance to capability, and values change from protective to productive. In addition, leaders and managers change from supervisors to coaches, and the organisational structure changes from hierarchical to flat. Last but not least, the executives change from scorekeepers to leaders [4]. This whole idea can only be seen as a major change from what exists in most companies today.

However, before we continue, we need to know what re-engineering is. Hammer and Champy define it as 'the fundamental rethinking and radical redesign of business processes to achieve dramatic improvements in critical contemporary measures of performance such as cost, quality, service and speed' [4]. They say that once the real work process has been re-engineered, the shape of the organisation and its structure needed to implement will become more apparent. The authors of the re-engineering concepts are very clear in stating that it is not changes to the organisation that are re-engineered, but the specific processes that make the product, project or service more applicable to the customer [4].

They also point out that they have found that most processes tend to be sequential, making them slow and clumsy [4]. What re-engineering introduces is the means to change the sequential parts and manufacture things in a more parallel manner, allowing for shorter time in process and a more logical approach to assembly and service.

The four requisite characteristics of re-engineering are:

1. the fundamental rethinking by the leaders of the company that focus on the existing process orientation in the organisation,
2. the new ambition of the participants to change the way things are done,
3. rule breaking by the leaders that does not accept the old traditions as they exist in the culture and
4. creative use of current Information technology, allowing the enabler characteristic to work in radically different ways.

These require the establishment of a selected process team that will look at the processes to make the changes and then dissolve when the work is done. The idea is to eliminate the non-value-added work that most processes include and to look for productive ways to exclude or add only productive applications to the overall work approach [4].

One of the major concepts accepted in this book has been the idea that any process or procedure used in the delivery of a project, product or service is open for review, suggestion and re-establishment as new and innovative. Any existing process or concept can be improved. Without question it is the stakeholder who has the key to this, with their knowledge of how the work is done. If it can be improved, it should be; if cost reduction can be brought to bear, it should take place. This is a key to re-engineering and it can be done every day in every role or job in the modern company. However, it has to be motivated by those who control the administration and the funds. Without motivation, acceptance and reward this process or step will not take place. This goes back to the previous step: reward is too important to be left to serendipity.

Let's review what we discovered in Chapter 2 about business processes re-engineering. Quality Process Magazine encouraged the following 'Golden Rules' in 1994:

1. organise by product,
2. redesign the process flow and
3. maximise the number of workgroups to meet the need.

Sixteen commandments were also recommended (see Chapter 2) [5]. The issue is that these requirements would necessitate an immense training or education programme. However, the key factor is that only through the use of value stream analysis will one really be able to conduct an efficient review of the processes and then eliminate the unnecessary or non-value-added activities. That is assuming the whole process is assessed, and not just a piece of it.

It is therefore suggested that the tool of choice for re-engineering should be value stream analysis [6]. This will enable the team looking at the process to assess the full effect of the value-added and non-value-added activities involved. It must be emphasised that when the process is completed and the analysis has been done, the team should also look at the whole process, as that this might be a subset of a larger process and any adjustments will affect the other connected processes in the overall system. Without this full assessment the effect of the changes may have no impact on the overall operation.

6.5   The fifth step

Leaders in today's organisations seem to think that as the company progresses, employees will automatically change to fit the requirements established, and as the company changes, their needs and capabilities will also change. What they don't realise is that the company requirements to a stakeholder, and those that they follow, are often those set by the culture in which they operate. That culture might be different from department to department, especially if the company has not taken the time to look at and develop the processes needed to instill its own new requirements. Therefore, it is incumbent upon leaders to be aware of the vision, mission, product development and objectives of the company in order to see where and how they must impose the required changes and develop a body of knowledge about what is and is not working.

It is suggested that the reader look again at the rewards Section 6.3 above. Again, these results are best found by conducting a value stream analysis to develop the positive and negative needs or activities for the group – what is of value and what is not – focusing especially on what they direct and lead. If a leader has not worked with their stakeholders to coach, mentor, teach and develop their skills regarding variation on all their processes for which they are responsible that concern the customer and the supplier, those stakeholders will be unaware of the events in the work group, especially where things have changed. That contact between the leader and the stakeholder works two ways: one is the detrimental demand on the stakeholder to produce as needed; the other is the feedback the leader gets from the stakeholder about shortcomings and needs (Chart 6.2).

1. Tolerated Innovation & Risk Taking

2. Attention to Detail & Reward System

3. Outcome Orientation

4. Impact of Leadership Decisions

5. Team Orientation

6. Company Atmosphere

7. Long-Term Perspective

Chart 6.2   Seven Characteristics of Organisational Culture [Source: From Conner, D.R. Managing at the Speed of Change. New York: Villard Books; 1994] (Chart 6.2). Note: See Appendix section for full page image

Leaders have to understand that their personal actions are watched at every movement by stakeholders. What they say is recorded in the minds of their followers but not necessarily followed through at first blush. The reason for this is often very simple: the stakeholder is trying to find out just how well the leader really believes what they have said. 'Walking the talk' sounds like a cliché; however, if leaders do not follow through on what they personally say, their word will be seen as being meaningless and without value. Once that happens, the rest of what one says will hold no meaning for any of the others in the group. That's how the company culture works; it spreads the bad word fast and the good word slowly. Ideas that are expected move especially slowly. They go slowly because the leader's word is always being tested. Once the leader understands this reality they are more able to get things done more effectively because they follow through on what is said and often repeated, and 'walk the talk' on a consistent basis.

It is essential that leaders provide all of their followers with a list of changes to be made to their processes – that is, the new approved adjustments to process and procedure from the CCB. They might not need this information right now, but they will in the future. When a stakeholder feels that they are a part of the group and are kept informed, they will be more forthcoming with information to the leader and more receptive when there is a need or requirement that must be changed or followed. It is trust in their leaders that most stakeholders hold most dear to their operations. When they lose this trust it becomes difficult for them to continue to do their job and their feeling of affiliation falls below that which is desired. That loss of affiliation will result in a desire to flee or leave and will set up a situation in which the stakeholder may be looking for a way out or a way to hide from attention in the scheme of things.

6.6   The sixth step

Where do we begin to identify the core competencies that make the company what it is? We begin with the products or services that are produced by that organisation. Generally the company is populated with people who are skilled in working on the specific core criteria that make it what it is. These core capabilities generally give us a clue as to what our core competencies are. How well do the leaders of the company know what the core competencies are – hopefully, all of them? How well have they been documented and communicated to the entire company? Often not at all. Are there processes and procedures that are used by the stakeholders to function within the confines of the policies and requirements and that of the company? If so, what are they and have they been documented, or do only a choice few stakeholders know what they are and others take direction from them as the quasi-leaders?

Once the core competencies have been identified it is incumbent upon the leadership to ensure that there are stakeholders who can do the job and fill the identified roles for each deliverable and work package. This is often done through knowledge of who can do what and how long it will take to accomplish that task, as well as the processes that go with it. If we don't know what the capabilities are, we are in a big world of hurt. It means that we have a job to do, but do not know who to assign the jobs to, and have no idea who can do what is necessary to accomplish the tasks required. Knowing the capability of the staff means that we know what the body of knowledge is for each role and can identify the people who can fill that role. There might be some discrepancies where all the skills required are not present, but knowing the body of knowledge enables us to determine the gaps for further coaching, mentoring, teaching or training.

Remember that a body of knowledge is made up of more than just the knowledge of the individuals. It consists of skills, abilities, knowledge, processes, methods, tools and experience. Sometimes it even includes certifications that a role might require. The same body of knowledge might have different criteria for a different role but be the same in its overall description of the requirements. That is, the criteria for a similar body of knowledge might have more critical requirements because of the severity or complication of the job that must be done. This helps us determine the capability of the roles and the jobs that are required to provide the deliverables or work packages.

6.7 The seventh step

In every organisation, the element of change must become the accepted and adopted concept in the company by all the stakeholders, managers and employees alike. Everything that is done must be evaluated with a questioning eye and the attitude as to what one can do to make the company processes operate better than they currently function. The questions have to be asked as to what can be changed, and how does one go about making those changes? A policy must be set in place. The role of the change control board (CCB) must be fully understood by everyone, and the CCB must be in place with appropriate policy and must be used by the company as a tool to improve.

So, what do we know about the nature of the change? Is there more for us to understand and pass on to the stakeholders under our mentorship? Collecting data regarding the change is an important activity, and communicating that information is just as important. What is the accepted process for making the change? Has the leadership taken the time to look at that process? And do we as a group understand the ramifications of that action? If there is no process and this is a random change requirement we have already lost the battle with those people who resist change altogether. The resisters will win out because of a lack of commitment and understanding about what the change will do for the overall operation.

As change is defined, it becomes both an important ingredient in the company's process improvement gains and the means to identify new and robust products, projects or services that the company provides. The important ingredient here is that item spoken of in past chapters: leaders retaining the opportunity to read and study new ideas. This can only happen if the leader has properly coached, trained or educated their workers to take care of the responsibilities they have as employees. The free time that this provides the leaders enables them to look at new ideas and read articles that describe new approaches, and becomes a platform for innovation.

'Innovation' is a whole new realm from which the leader can identify and support new ideas, products or services; yet without the time available to study they might not have known of them at all. Now the issue of resilience to change really comes to the forefront. With resilience, the leaders can now look at what is happening around them and discover what needs to be done to enter a new realm of endeavour or provide a service long awaited by the customer at large. With this new time available, the leader needs to audit their information intake to the kinds of things that appear to be important and to develop observational skills that look at these things from a different perspective. As they ask questions and watch the trends of their business, the readings and ideas that are presented through various media will start to develop new ideas and approaches that they can use. Remember that the innovators don't create the wave of innovation; they merely amplify it and help to popularise the ideas or services for others to use. Chart 6.3 emphasises the prime management functions of leadership.

Being able to spot opportunities is a change-oriented skill. It goes arm-in-arm with the potential success of the company and the leader's ability to be resilient when faced with the adverse conditions that often accompany change and resistance to it. Resilience requires the leader to observe the trends and find a way to deal with them. With that comes the requirement to search for solutions in the face of adversity while realising that some of what you as a leader do might appeal to others as well [10]. What unexpected successes have you had and what might they mean in the scheme of things? How are you dealing with trends that are running opposite to your normal approach? Sometimes the worst events are flush with the greatest ideas that you might not have thought of at all. Where were you and what was happening when you noticed the handwriting on the wall?

We cannot ignore our competitors and just say they don't exist. They do, and they often have some fantastic ideas. Sometimes they come up with the most ingenious ideas. How are you dealing with them and how resilient do you feel when you have to come up with an idea that will put you back in the running against the competition. Breakthrough ideas occur when you are up against the wall, under pressure and searching for new opportunities. Innovators are prepared to act because they do not fear change. If you know what your people are capable of and can count on them to work with you that also provide more than a little help [10]. For this reason I wanted to remind the reader of the process required to verify the capability of your workers and the means required to reduce the gaps by coaching, teaching and mentoring in their capability arena. See Chart 6.4 on the body of knowledge development.

- A Primary Focus Must be on Teaching & Coaching to:
  - Control Processes & Keep Work Distributed
  - Satisfy the Customer
- Develop Long-range Customer Driven Plans
  - Involving All Employee Inputs
- Devote Time to Studying, Thinking & Learning
  - About Bold New Innovations
  - Concern for the Success of the Company

Chart 6.3   The Prime Management Function is Leadership. Note: See Appendix section for full page image

It should not be a surprise that the most important element of change is the ability of the leader to be resilient. As demonstrated by Chart 6.5, many items impact on the leader, who must bring change to the company and understanding to those who serve as stakeholders and followers in executing that change [7]. No matter how astute the person, without resilience the potential of the change and the ability to execute it will not take place.

6.8 The eighth step

Accountability is a must. Without true accountability and assignment of responsibility for the financial aspects of the project, programme or service, the operation upon which the company is dependent will fail due to inattention or a lack of concern from uninvolved stakeholders. Leadership and its attendant parts require that responsibility, accountability and the attendant reporting are assigned to finances. The measures used to determined appropriate attention to detail are up to the management and leadership, but they should be arranged and required of the responsible resources. This is where the leader's ability to ask questions and track the events really counts.

Accountability for the leaders and the team, whether they are peer or programme personnel, begins with them accepting periodic status and control meetings. These meetings enable the leaders and members of the team to hear the same information simultaneously; they enable them to monitor and control activities, identify problems and establish a consensus on the solutions to issues. They also allow the members to assign future activities and provide feedback on team performance since the last meeting. To be functional the team and the leadership need to review the accomplishment and any concerns with a view to self-evaluation and adjustment of their operations.

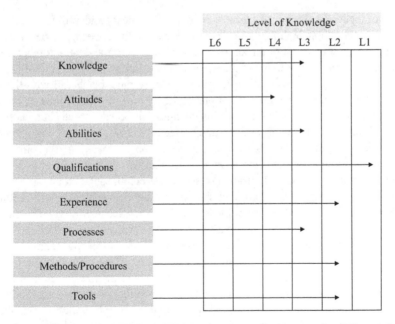

Chart 6.4   The Body of Knowledge. Note: See Appendix section for full page image

This need is an opportunity to bring in the benchmarking tools and look at the performance of the competitors as well as other teams in the company. Comparing what can be accomplished with what has been accomplished is an important accountability step for the leader and the team. Without this comparison, the team is just looking at its own efficiencies and deficiencies. Comparison with others allows the team to see new approaches and problems that they might have created but were unaware of. After such a review the leader and team members should set out to track their existing goals again and compare them to what they might be doing with the new information they have acquired.

Without question, top-level management should be brought in periodically to see if its corporate goals are being met and how the accomplishments compare to the overall expectations of the company. Perception is a nasty enemy: if it is going against the team it can devastate the project. To set the perception right and deal with any misconceptions, the leader and the team must have the top-level management involved in a periodic review at which the air can be cleared and the record set straight. Any misconceptions or perceptions on behalf of the leadership or management will arise at this time and be subject to discussion and clarification. It is the duty and responsibility of the project leader and the team to deliver the product with realistic specifications and requirements within schedule and cost. Without this guarantee, the effort is a total waste of time.

6.9   The ninth step

The role of leadership in managing the sub-contractors must be identified, written as a document, understood by all involved and followed. Assurances must be made that the requirements for the work packages will be shared with the contractors as well as that the changes that occur to the requirements, configuration and the build of the total work package will go forward. Configuration changes must be shared with the contractors as well so that they are not left out of the loop of new directions determined by requirements changes and configuration changes. The cost of these changes should be worked out in some detail so that sub-contractors do not feel they are absorbing costs that are not due to their actions but those of the company. At the same time, open the door to subcontractor input on changes that they might see in the development of their part of the project, product or service. When the subcontractor is brought in, they become a part of the team and should be treated as such. The subcontractor may then be more willing to work out lesser costs as they see new ways of getting the work done or savings that might be passed on to the company and the client.

Subcontractor leadership needs to be assigned from within the team. The best leader is the individual who has the specific role or job on the team that requires the subcontractor to perform as a member of the team. It is incumbent upon the team leadership to make this role a clear and understood job that someone must undertake and fulfill in the best and most accepted way possible, by following the established and accepted processes and communicating with subcontractor and team members alike. Does the subcontractor understand the requirements for the work package they have been assigned? Are they on board with the configuration, cost and schedule for completing this work package? Are there rewards and penalties in place that make it clear to the subcontractor how the work package and its resultant product will be assessed, evaluated and controlled? If these guidelines are not understood, the result

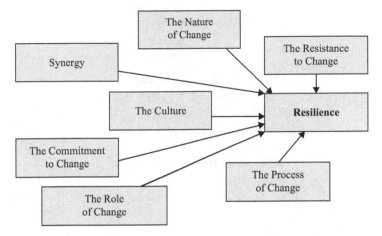

Chart 6.5   D.R. Conner Says That the Elements of Change Are Best Met with Resilience [Source: From Conner, D.R. Managing at the Speed of Change. New York: Villard Books; 1994 [7]]. Note: See Appendix section for full page image

may not be what the team is looking for and the work package may not be what the company wants or needs. If the subcontractor understands that they are an integral part of the team, and that they will be rewarded for any savings and improvements made, they will want to play that role. It needs to be clear to them how they can do that. There should be a role description with the body of knowledge that describes how they are able to contribute to improving the work package.

6.10   The tenth step

What type of tracking and oversight has been put in place to oversee the entire project, product or service? Without this part of the planning process in place and with a clearly delineated policy, there will be little hope for success in the long run. The policy should be clearly stated, with the required processes in place and the procedure worked through with all of the associated stakeholders [8].

Have we done an effective planning job? Is enough known about the project, product or service to allow us to determine adequate requirements that will enable us to do a good job of configuration and production on the work package?

The most important part of the planning process is the establishment of the engineering development plan (Chart 6.6). This should be used to track the activities of the work package. The status of all the aspects of the plan should be communicated to the members of the team and up to the next level, where the package will be integrated. Inherent in the plan are the verification processes that will be used to track and oversee the operations. As the work progresses, accomplishments should be communicated, and when milestones are completed this should also be passed on. All affected stakeholders should be aware of the status of the package,

1. Create the Operational Definition for the Company
   - What is the Product, the Vision & the Mission?
2. Develop a Risk Management Strategy
3. Build a Useable Work Breakdown Structure
4. IDs for the Task Relationships
5. Estimate the Work Packages (Time & Cost)
6. Calculate the Initial Schedules
7. Assign and Level the Available Resources

Chart 6.6   The First Steps of a Real Plan. Note: See Appendix section for full page image

especially any revisions, refinements or changes that have been approved. Something that is often forgotten are commitments by the leader to others outside the team, especially top-level management. Changes that will affect these commitments must be communicated as soon as the known adjustments have been applied. Where tracking and oversight is the rule, these items will be known quickly and reacted to just as quickly.

Where cost and schedule are affected and tracked, a communication to everyone is more than suggested, it is required. It should be dealt with quickly, especially if there is an increase in either. The corrective action should also be known by all affected. To enable the leader and management to know what is being measured and tracked, it is wise for a tool to be applied and put in place where it can be used in the communication process. Microsoft Project Management is a package that should be looked at carefully and might be used as the key measurement tool for schedule control and assessment. If the reader has a different package that serves as a good Gantt Chart measurement tool, it should be used.

With these tools to hand, periodic meetings with the team and the affected players should be scheduled. Weekly meetings are suggested. If possible, peer review teams should also be established because they bring a different viewpoint to the project. Peer review teams are made up of people who are totally disconnected from the work package itself but have the expertise to be able to look at all the activities and can comment on the wisdom or lack thereof and make suggestions that might improve the process and activities for the team. The peer review team would be brought in periodically to review the schedule, cost and activities whenever the feedback would be of benefit to the actual team [9].

Periodic formal reviews should also be established at meaningful points in the schedule to ensure that progress is being made. These should be conducted with the end user where possible and all affected groups in the organisation. Top management should also be invited to hear what is said about the project, product, service plan or package. Earlier in the book, it was stated that there were key processes that are recommended in the third stage of the CMM that would benefit the reader. Peer review is one of these, alongside training. Use of these processes will help the leader and the organisation conduct meaningful formal reviews.

The use of level three key processes is not restricted to when one is focusing on the level two processes towards becoming a mature organisation. Of course level two is of great importance to the engineering function, but the level three processes that support and help build a stronger organisation are the best use of time and effort. It is for this reason that the author suggests that the training and peer review key processes are considered and used.

To review, here are some of the key processes and issues that the leader should be aware of and exercise in their business environment. These are not given in any specific order, but they support the ideas presented in this book.

The leader should be strategically oriented to understand the overall corporate mission, know the key strategies and be committed to the success of the company. They should be aware of the business conditions and how they link with the strategy, and be monitoring and guiding the trends and opportunities for maximum success.

The leader should also be aware of the income issues – how the finances affect the other organisations and where their use of technology has been the most successful. Alongside that a thorough understanding of the competition and the organisation's customers should be demonstrated in their daily application of these skills and abilities. A good understanding of the company culture provides knowledge of where and how to introduce the required changes and adjustments to the daily operations, as well as how to be the resilient component in the mix [11].

In addition, the leader is expected to exhibit a high standard of performance to the other stakeholders – to walk the talk – be an effective listener, an action type, and trusted by their staff. The effective leader is expected to be able to stretch their stakeholders, thereby fulfilling their potential, and can identify the most capable performers. That same leader is expected to communicate effectively both verbally and in writing, manage conflict and change, and above all prepare their successors through coaching, mentoring and teaching. This adds up to a lot of requirements for the most capable leaders, but it is what is expected. With proper attention to the key issues and ideas presented in this book, one should find the tools are there to be used by that capable leader, and then some.

Questions for the reader

1. Do you know what your company's core or key technologies are?
2. If you know what your core or key technologies are and have hired stakeholders to support them, does your company have a reputation in that field?
3. How do Maxwell's Laws fit into your way of thinking about your company's reputation?
4. What type of reward and punishment system does your company have? Have they ever given an indication that they are open to considering a new approach?

5.   Is your company's reward and punishment system focused on the objectives, goals and vision? Is the team rewarded or punished accordingly for meeting or not meeting the company's focused items?

6.   Does your company reward the top-level management before the stakeholders? Is there a Management Incentive Compensation Plan and no reward system for the stakeholder? What type of stakeholder reward system is in place?

7.   How does your company ensure the 'trust factor' in its daily operations? How would you improve that factor if you were the leader of a project?

8.   What items does the Sibson Study brings to the table for a leader?

9.   How would you apply Sibson Study information to your company?

10. Would the EVP have any effect on your company? Explain.

11. What is the biggest roadblock to a company conducting a business process re-engineering project?

12. What happens to an organisation that has been re-engineered?

13. How well have the core competencies been documented and communicated to the entire company?

14. Are there processes and procedures that are used by the stakeholders to function within the confines of the core processes and that of your company?

15. Knowing capability means that your organisation knows what the body of knowledge is for each role and can identify the people who can do that job who are filling those roles. Do you believe your organisation is aware of this? Are they able to assign the correct people to the needed roles?

16. What are the parts of the body of knowledge? Can you name and define them?

17. How do you determine the various levels to each item in the body of knowledge?

18. How well is change accepted in your company? What is the company policy regarding change?

19. Why is resilience so important to an individuals' capability to deal with change?

20. What does financial accountability have to do with leadership? How is financial accountability done in your company? Do you feel it is fairly done?

21. Does the sub-contractor understand the requirements for your work packages when they are let out?

22. Are your subcontractors onboard with the configuration, the cost and the schedule for completion?

23. Are there rewards and penalties in place that make it clear to your sub-contractors about how the work packages will be assessed, evaluated and controlled?

24. How do you ensure that the tracking and oversight process is conducted?

25. Are there tools that can be used to factor in the tracking and oversight requirement?

26. How is the process of peer review utilised to ensure effective leadership and management?

References

1. Maxwell, J.C. *The 21 Irrefutable Laws of Leadership*. 10th edn. Thomas Nelson Publishing; 2007
2. Guderian, B. *Leadership Development and the Role of Continuing Education*. Presented to graduating class of the ELITE Program; Tulsa, OK, 2010, p. 2
3. Conlon, R., Insler, D., Kochanski, J. *Rewards of Work Study*. Sibson Consulting; 2009. Available from http://sibson.com/publications-and-resources/surveys-studies/?id=252 [Accessed 5 June 2013]
4. Hammer, M., Champy, J. *Reengineering the Corporation: A Manifesto for Business Revolution*. New York: Harper Collins Publishers; 1993
5. Quality Process Magazine. 'The Golden Rules and Commandments of Business Process Re-Engineering'. *Quality Process Magazine*. December 1994
6. Keyte, B., Locher, D. *The Complete Lean Enterprise, Value Stream Mapping for Administrative and Office Practices*, New York: Productivity Press; 2004, pp. 6–8
7. Conner, D.R. *Managing at the Speed of Change*. New York: Villard Books; 1994
8. Lareau, W. *American Samurai: A Warrior for the Coming Dark Ages of American Business*. New York: Warner Books; 1992
9. Humphrey, W.S. 'Characterizing the Software Process: A Maturity Framework'. *IEEE Software*. 1988;5(2): 73–9
10. Waitley, D.E., Tucker, R.B. 'How to think like an Innovator'. *The Futurist*. May–June 1987, pp. 9–15
11. Vicere, A.A., Bitner, S.W., Freeman, V.T. *Management Skills Assessment*, Penn State Executive Programs, 1985
12. Vicere, A.A. 'Breaking the Mold: Strategies for Leadership'. *Personnel Journal*. May 1987, pp. 67–78

Chapter 7
Individual capability

Figure 7.1   Martin bombers at Schofield Army Air Base (1934)

There are three components to individual capability:

1. the self, or ability to function with meaningful people who have leadership skills,
2. process leadership skills, or 'operational leadership' and
3. the skills and ability to work within the organisation and construct successful organisational leadership policies.

By reiterating those skills illustrated in earlier chapters, and adding new categories, this chapter will help the reader develop the necessary skills in these three components. It will also define capability as seen by the organisation from many viewpoints (Chart 7.1).

Capability is an important concept and must be carefully assessed when trying to improve the operation of a company. Individual capability is the people part of the operation. It includes the skills, abilities, knowledge and experience that people bring to the organisation when they are hired. When we hire someone, we are looking for the specific skills required for the role or job. As well as that is the ability to apply those skills and knowledge, focusing on the how and why that explains how things are done. Experience must also be a key consideration, as knowledge and ability are often learnt from experience. This can be discussed in the hiring interview, which is a chance for the interviewer to discover what the prospective applicant really knows and doesn't know (Chart 7.1).

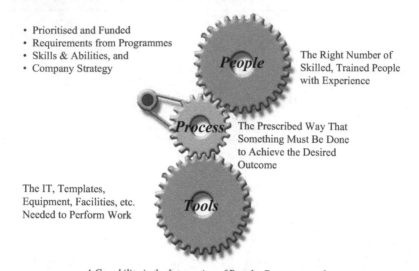

- Prioritised and Funded
- Requirements from Programmes
- Skills & Abilities, and
- Company Strategy

The Right Number of
Skilled, Trained People
with Experience

The Prescribed Way That
Something Must Be Done
to Achieve the Desired
Outcome

The IT, Templates,
Equipment, Facilities, etc.
Needed to Perform Work

*A Capability is the Integration of People, Processes and
Tools Working Together to Create a Valued Result*

Chart 7.1   What is Capability? Note: See Appendix section for full page image

It is critical that the hiring leader or manager has done their homework before the actual interview takes place. This should involve a delineation of the skills required to do the job and fill the role, with the associated abilities, knowledge and experience that are needed. Skills come in varying levels, from basic to highly skilled. It is quite important that these be determined for each job and stated in the job or role description before the interview process begins. A job role and its associated levels of skills, knowledge and abilities should be established and made available to Human Resources so that they can do the necessary advertising and review the resumés of those who apply. Establishing the levels of skills, abilities and knowledge might be seen as a function of establishing the body of knowledge for the role.

Looking at Chart 7.2 we can see that a competency structure can be used to develop the requirements of a company role or job. Once the need and requirements are determined, the capability for that role has been determined. Notice that the 'body of knowledge' incorporates more that just the skills, knowledge and abilities (SKAs). However, for now it enables us to begin determining the needs for this role.

Using the difficulty levels of 1 through 5 as an example, we can now determine the skill level, knowledge level and ability level we will expect from the person we hire. The level of difficulty does not have to be the same for each of these concepts; it may vary from the most difficult to the least difficult depending upon the needs of the role or job. If we look at Chart 7.3 we can see that any one characteristic might have a different difficulty rating for each concept. This means that the hiring leader is able to categorise capabilities presented by the candidates that might be grossly different, and that in the assessment might have a desirability level that is either acceptable or unacceptable depending on the specific need or characteristic requirement.

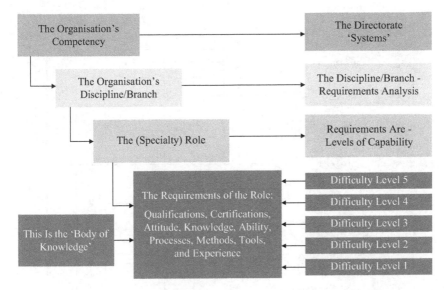

Chart 7.2   The Competency Structure. Note: See Appendix section for full page image

Let us look at an example. If we were hiring an experienced systems engineer, we would want that person to be very knowledgeable about requirements and have a high level of capability in determining them for any type of product or service. They should also be knowledgeable of configuration management techniques, but because configuration is the responsibility of another person they need not have as high a knowledge base in that concept – understanding is as important here as application. Quality may also be of high importance, but not as high requirements. We might set quality at a medium level. The issue of processes would be higher, as allowing others to conduct their processes under the systems engineer's guidance is important and highly desirable. With this thought in mind, we look for the candidate's ability to guide the stakeholder in each case with which they deal, but at the same time they must be able to hold them accountable for the results and effectiveness of the process operation of which they are in charge. One might also be concerned with the individual applicant's ability to guide others to develop their procedures that support the processes.

Now we have a measure that allows us to assess the candidates for a specific role or job that we are offering. This guide helps us develop the requirements for the role or job and allows us to assess the applicants for it. I am sure the reader knows that it is not possible to find or specify all the characteristics required. Therefore, while the interviewer is doing their job they can assess the candidate's interest in the job, how they have approached similar situations and their willingness to grow by developing in the role over time. Knowing all of the characteristics required of the role allows the candidate and the hiring reviewer to determine any gaps in the capabilities of this person. The fewer gaps that are

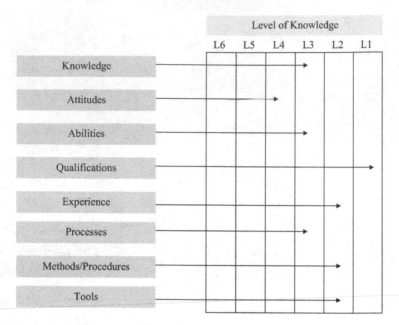

Chart 7.3   The Body of Knowledge. Note: See Appendix section for full page image

identified, the better they fit for the job. However, where there is a very positive attitude, a willingness to grow and a shortage of gaps, the hiring manager can see that this may be a fit, especially if the role is not in a critical area. The candidate will also see meaningfulness in the role that they can fill with a determination to develop over time while in place. This gap analysis can be used to determine the training, education or coaching that will be required if the candidate is hired.

Hopefully the reader can see how important it is to determine the specific SKAs and level for identifying the best fit employee for the job or role. It is just as important that the leader or manager sees the employee's ability to adapt to the requirements to develop based on the qualifications brought to the table by the search this may be brought, by the candidate and by the HR organisation for the qualified applicant. Where there are gaps? Who is the best fit and how much work will have to be done to eventually get the chosen candidate to the required level for the role? Knowing where the gaps are and being able to train or educate the employee to reduce them makes the company more successful.

7.1   Repeatability: the key process for leaders and their employees

The key processes for leaders and their employees are at level two of the Capability Maturity Model (CMM) Version 1.1. This is the repeatability or project management level, and includes project planning, requirements management, configuration management, quality assurance, subcontractor management, project tracking and oversight. Each of these processes is a key concept in the leader's toolbox for assuring that every project will be similar in results and completion to any other that they might take on. Two key parts of the ELITE Leadership Model are shown in Chart 7.4, in which 'repeatability is determined by self and people-oriented leadership skills'.

- Self Awareness
  Development
  - Self Awareness Skills
    - Self Awareness Skills
      of Effective People
  - Management Skills
  - Social Awareness
    Skills
  - Good Relationship
    Development Skills

- People Leadership
  Development
  - Effective Team
    Building
  - People Development
    - Coaching
    - Mentoring
    - Teaching
  - Motivation Skills
  - Functional Courage

Chart 7.4   Self Awareness & People Leadership. Note: See Appendix section for full page image

Before determining the requirements, two skills from the ELITE Leadership Model should be examined carefully. The first is self-awareness development, and the second is people leadership development [1]. Being aware of one's capabilities is essential when beginning a project, product or service. If a leader is able to understand themselves and use the skills that they have developed, they will be able to follow through with the other stakeholders working beside them to determine what is needed. Knowledge of the self helps development of the team, which is a people leadership skill.

It must be obvious by now that the body of knowledge is a key concept in developing the capability and long-term ability of a company to stay current with the technology and function of the roles it has assigned to stakeholders. When a company stops keeping up with the changing technology and changing environment that affect its roles, jobs stagnate and the company is unable to keep up with the needs of its customers or the environment. The body of knowledge determines the requirements for the job or role. However, they are also determined from examination of the company's core competencies and the processes, tools and methods required to get the job done. That is to say, if we know what core competencies are required to satisfy the company's vision and mission we can easily identify the roles that will be required to expedite those jobs and complete the project, product or service.

Each of the core competencies will have a set of required roles. Within each role will be the set of skills, abilities, knowledge, processes, tools, methods and experience that make up the body of knowledge required to execute the role. This is

1. Identify the Critical Technologies & Processes Used by the Company

2. Identify the Baseline Roles Played by Staff to Support the Critical Technologies & Processes

3. Determine 'Body of Knowledge' (Skills, Knowledge, Attitudes, Abilities, Processes, Methods, Tools & Certifications) Required to Support the Roles

4. Identify the Capability of Each Role, Each Process & Each Technology – Where Are the Gaps?

5. Determine the Number of Resources Who Have These Capabilities

6. Determine the Capability Over Time with the Resources That Are Available, Developing & Planning to Retire

Chart 7.5   Eleven Points for Knowledge Development. Note: See Appendix section for full page image

why we have to start with the core competencies. It is also not uncommon for a stakeholder to have more than one role. A complete analysis of the needs is a must for any company (Chart 7.5).

Once the core competencies (technologies and processes) are known we can begin to identify the baseline roles required and played by the staff to support these capabilities. Now that we know the roles and the people who are filling them it becomes important to determine the body of knowledge required to support these roles. Imagine that you don't have the cooperation of the people in those roles. What kind of problems are you going to have getting the information required about their capabilities? If you are not a leader, and cannot gain the support of your stakeholders, you will not be able to determine what capabilities the stakeholders have and you will definitely not be able to assist the specific role in developing their capability gaps.

The fifth key point is to determine the number of resources that you have in any one capability requirement. This knowledge might be helpful when you have too many stakeholders and need to eliminate some of the redundant ones. However, keep in mind that you are dealing with a culture and protecting roles and other individuals may just be something that you have to overcome. Leadership can never be forgotten when filling the role of manager or supervisor; it might carry some baggage with it.

Point number 6 can generate a lot of resistance from the leader's supporters. If one is not truly seen as a provider of development, affiliation skills and their nuances, who has an understanding of the organisation, this data will be very difficult to gather. Cooperation will be much more forthcoming if the leader is developing stakeholders through mentoring, coaching and teaching. Roles and jobs that are threatened will come under a form of protection much more quickly than those that are not threatened. When employees see their leader as someone who develops them and continues to look out for their growth and advancement, they will be more welcoming of any questions that have to be asked. Most stakeholders want to be able to retire with dignity – when they see that their leader is making that possible as well as supporting others in the group they will support the leader's questions and provide the necessary answers. In many cases one will find that the stakeholder might volunteer to be a mentor or coach for a younger more inexperienced employee.

7. Identify Methods in Place to Maintain Capabilities
   – Mentoring, Cross-training & Internships
   – Programmes, Courses of Study & Re-Training (In & Outside)
8. Establish Policies & Publish to Ensure Action for Critical Needs
   – Performance & Gap Assessment, Quarterly Review
   – Management Development, Skill & Knowledge Training
9. Establish Change Review Board – Meets Monthly to Enact Change to Policy & Process
10. Eliminate Capability Gaps in Training Workforce or Hire New Employees That Have the Required Skills
11. Maintain Education & Training Programs for New Skills, Changes in Operations & Gap Reduction

Chart 7.6   Eleven Points for Knowledge Development (cont.). Note: See Appendix section for full page image

If the policies are in place they should be published and discussed by all involved. How do they work for the betterment of the stakeholder? To reiterate, we are most concerned as a company that the critical or core competencies are supported by trained and capable employees who are either able or developing to become everything that they can be for the roles that they fill. If the policies don't support this, there is a big problem that needs to be fixed. This can be done by having quarterly one-on-one meetings with the stakeholders to gather input. The ability to communicate and do it well with all involved is of the utmost importance. Where leaders lack the skills that are required, they may need more training themselves. Now we can see why the idea of knowing the self, as identified in the ELITE Leadership Model, is so important. Where there are gaps, one must work to develop that skill or ability to identify the tools and methods to be successful (Chart 7.6).

Does the policy include the establishment of a change control board (CCB), or does one already exist? If it doesn't, then the policy should be developed, supported at all levels, especially top management, and put into place to support the changes to requirements, configuration, processes and plans. Without the policy and the placement of the best people in the company on the CCB, the appropriate changes cannot occur to keep the stakeholders happy and the customers satisfied. In addition there will be no changes that could save the company money and improve the quality of its product, project or service.

When one thinks about it, the last two items on the list are really givens. Once you have identified the gaps that exist, can you coach, mentor or train for that requirement? If many of the stakeholders in those positions are close to retirement and even ready to leave, you hire the best available with the skills that are required and coach, mentor, or train for the remaining gaps. But it is incumbent upon the leadership to make sure that the stakeholders are aware that development is important and available to them when new skills are required, as well as when gaps are recognised and in need of development.

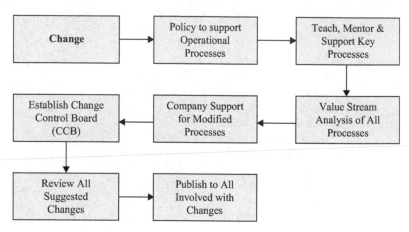

Chart 7.7   Change, Policy & the CCB. Note: See Appendix section for full page image

Chart 7.7 shows a flow chart that can be used to illustrate the use of the CCB in any company that supports this type of policy. First you have to recognise that change is required, and what that change will be. The next step is to develop a policy that will support the change and bring it to the attention of the CCB. If a value stream analysis is required it should be done. The result should then be brought to the attention of the CCB. If they agree, then the changed process should be published for all to see and use.

7.2   Repeatability: focus on project and programme management

When focusing on project or programme management, one is dealing with the fundamental issues identified by Watts Humphrey in the early days of the Software Engineering Institute. These are:

1.   requirements management,
2.   configuration management,
3.   project planning,
4.   quality assurance,

5.   subcontractor management and
6.   project tracking and oversight.

Chart 7.8 gives the complete CMM [2]. The example illustrated is CMM Version 1.1, and shows the key processes for all of the five stages of the model.

The following is a reiteration of the stage two key processes. The second stage, repeatability, is Humphrey's assessment of the needs and requirements of programme or project management that includes basic project management processes such as tracking, cost, schedule and engineering functionality. When in place these process disciplines allow operations to repeat earlier successes through the use of similar applications. Planning and managing a successful project is based on the experience of similar events. By institutionalising the prescribed management processes, organisations can repeat and implement projects that might differ only in the image itself.

Organisations that have installed the required processes and controls are able to plan, establish functional requirements and configure them appropriately. They are also able to manage their subcontractors, conduct effective quality reviews and track costs, schedule and functionality. The requirements and work products are appropriately baselined and their integrity is controlled. Stage or level two is categorised on the CMM as a strong capability because it involves disciplined planning and tracking with stable oversight and applications allowing for effective and efficient performance on the required project, programme or service.

A good analogy to illustrate the effectiveness of institutionalising the stage two processes is that of a typical engineering project. Established management controls allow the leader to look into the project – which might be the building of a succession of black boxes – and include things like a set of programme milestones. Even though the leader may not be able to see the details of what is happening in all of the boxes, the processes and their checkpoints can confirm that the process is working, and when it is not the leader can react accordingly.

7.2.1   Requirements management

The process of requirements management usually begins with what might be called evolutionary prototyping. This is where the team and IPT or IPDT group gets together to identify the structure of the prototype to be built as a project, product or service. Over time the team performs a number of walkthroughs with the stakeholders to elicit and validate the development of the necessary requirements. They use increasingly specific criteria to attempt to reduce ambiguity and incomplete information. The problem with this process is often that the team overlooks specific risks that must be mitigated and do not realise their presence until a later time, often after the baseline has been set. Managing risks has to be high on the priority list when doing the requirements, beginning with the identification, analysis and resolution or mitigation of need and risk. This is why risk mitigation and identification are at the very beginning of the process on the WBS model. A good example of a system used to regulate and control the risks is called the 'go/no go gate model'. This identifies specific gates, and whenever the assessment of requirements reaches one of those gates the team makes a decision to go or not go based on the analysis of the requirements. Is it obvious that risk management must be an ongoing process in requirements determination? In many cases it is not. For this reason it is suggested that the team include a systems engineer working with them on the requirements and the risks. This will allow the team to accomplish two things at once, and to do so with the effectiveness expected by management.

Requirements management and the development of the requirements are often thought to be an agreement by all on the technical and non-technical necessities of the project. It is considered the basis for the estimation, planning, performance assessment and tracking of the activities for the event. Incorporating these items into the overall plan is a necessity.

Risk must be part of the identification of the requirements before anything can be done on the project, product or service. That is to say, if we do this, then this might occur. Or if we build the process in this way then that might occur. Risk management is a must in the process. This can be seen on Chart 7.9 for developing the WBS [3]. Again, this can be accomplished by having a well-qualified systems engineer on the team.

As the team and stakeholders develop the requirements, the architectural design, sub-systems, modules, random specifications, interface design and manufacture should all be reviewed. First article (i.e. the very first prototype item to be built, establishing the first baseline for the requirements and configuration of the product) prototype design is a skill to be encouraged so that team members can re-evaluate what they have set as requirements and the potential baseline. Although this might be the first baseline, it should not be the last. The prototype and its output should be tested and evaluated for customer and stakeholder satisfaction against the criteria from the requirements and visualised baseline. Requirements analysis is a skill that should be developed by those working on a project, product or service development team.

Another factor to include is the presence of a systems engineer on the team, as noted earlier. Systems engineers are well-trained in the development of requirements, risk, mitigation and baselines, and understand the methods that should be exercised to develop effective first article needs. Using their skills and knowledge adds to the team's capability and overall productivity.

Managing the project, product or service at the system level is of the greatest importance. One must be aware of the customer needs, the organisational operation concepts and how the mission analysis affects the product. The requirements analysis must look at all the performance level expectations, the interfaces, correctness and the baselines. When deriving the requirements we need to understand what we are allocating to the teams and their operational requirements. Last but not least are the verification and the methods used to do it. What have we set as the acceptance criteria and evidence that we really have closure? When looking at the team's role we should also have verified inputs such as the performance requirements, products and constraints. When the team derives the requirements for their work packages, establishes their verification methods and validates the products we should be ready to verify from the company's point of view. This can be viewed in the flow diagram in Chart 7.10.

	Level	Focus	Key Process Areas	Result
Improve	**Optimising** 5	Continuous Improvement	Process Change Management Technology Change Management Defect Prevention	Productivity & Quality
Control	**Managed** 4	Product and Process Quality	Quality Management Quantitative Process Management	
Defined	**Defined** 3	Engineering Process	Organisation Process Focus Organisation Process Definition Peer Reviews Training Programme Inter-group Coordination Product Engineering Integrated Management	
	Repeatable 2	Project Management	Requirements Management Project Planning Project Tracking & Oversight Subcontract Management Quality Assurance Configuration Management	R I S K
	Initial 1	Heroes		

*Improvement initiatives **must increase market share**
and/or profitability in order **to have business value!***

**Version 1.1
C-SEI/CMU**

*Chart 7.8   A Process Management Model ... the Capability Maturity Model
[Source: Image adapted with permission from Zubrow, D. 'Putting
"M" in the Model: Measurement in CMMI'. The Software Engineering
Institute Conference, Carnegie Mellon University, 2007]. Note: See
Appendix section for full page image*

7.2.2 Configuration management

Configuration management is the leader's ability to see how something will be assembled, whether that be a project, product or service. That configuration will depend upon the team's ability to communicate with the leaders of the project and company for the development of an understanding of the operational definition established for the product, project or service. Understanding how something goes together is also founded in the understanding of how it has been defined and the requirements that have been developed for construction of the prototype or first article. Configuration management allows the leader and the members of the team to control specific work products or packages where changes may be made to the product, project or service itself. Effective configuration management means that the leader can inform all the affected groups and individuals of the status, change or baseline adjustments. This sort of information allows for more effective tracking and oversight of the work packages as production continues. The subcontractor also benefits from this data and the feedback provided.

The author recommends that a system for controlling and maintaining the engineering baselines be put into some sort of library or recording storage operation.

Systematically controlling and auditing the baselines and configuration functions is of the utmost importance. There is no question that this function should be planned and controlled, giving everyone involved access to the maintained data. Allowances to change any of this data should be reserved for the CCB and leadership combined. The CCB and the leadership should have the authority to manage the baselines and establish changes as required and approved. The groups that should be notified when changes are approved are: quality assurance, all leaders and managers, the manufacturing or operations organisations developing the product, systems engineering, contract management and subcontractors. All changes and notifications should be documented and recorded in the database or library, whichever is the case.

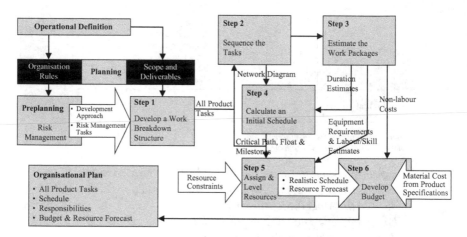

Chart 7.9   Developing the Work Breakdown Structure (WBS): Decision Making [Source: Reprinted with permission from Verzuh, E. The Fast Forward MBA. 2nd edn. New York: John Wiley and Sons; 2005]. Note: See Appendix section for full page image

The library or storage and retrieval location should be available to all parties of the project, product or service. It should provide for the sharing and transfer of information between all affected groups using the standards established and the updates available as necessary. The lifecycle of the product, project or service begins and ends with the documented requirements. Therefore, the storage and retrieval of all the established data is a given. It cannot be said more emphatically that the primary product of any engineering project is the documentation. This allows for the replication and repeatability of the manufacturing process and the successful building of the first article and those that follow [4].

Based on this knowledge, it can only be said that through frequent testing and experience, engineering firms have realised that correcting errors is an exponential job if the requirements are not clearly understood, are not applied correctly and if changes occur without knowledge being transmitted to those affected by the change. The author has seen over time the results of errors and corrective actions that have resulted because of others on the team feeling that they had the correct approach and did not bother to let others know, or just plainly did not take the required steps to get changes approved where it was necessary to do so.

Included in all of this is the importance of maintaining a concurrent approach to both the designs and the processes being applied. Without an understanding of the need for concurrent application, one might not realise that design and process are independent applications. So it is important to emphasise the need for the team approach and use of the concurrent skills as each develops. Some sort of requirement need would aid team members in this development; they will operate more efficiently if they know that all the work is requirement driven, and that this data is retrievable and kept in storage provided by configuration management. The highest level of integrity is only achieved when the stakeholders are validating the documentation through their actions [4]. Such an understanding between the organisation and the stakeholders ensures that as changes occur and approval is given, the affected participants are aware of the information and able to acquire the changes. This ensures a robust organisation. As explained in earlier chapters, organisations that are not afraid of change and are willing to deal with it in a robust fashion are often the most successful.

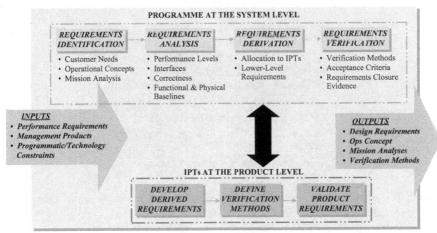

Chart 7.10   Manage Requirements. Note: See Appendix section for full page image

7.2.3   Project planning

Planning begins with a statement of work (SOW) developed by the team in charge of the work package. This defines the constraints, goals and boundaries of the specific project. From the WBS and the SOW, steps are taken to determine the elements in the work packages, the required resources, a potential schedule, the commitments and the associated risks. The project, product or service plan should provide a basis for performing and managing the work package activities, and should commit to those packages according to the available resources, the constraints identified and the capabilities of the team members. The operations to be executed for all the work packages are: the capabilities of the engineering group, development of the estimates, systems engineering, quality assurance and test, contract, configuration management and documentation support. This process can be viewed on Chart 7.11.

At this point the leader of the project is responsible for negotiating the commitments and developing the specific plan for completion. It cannot be said more emphatically that it is not solely the duty of the leader to develop this plan or make the commitments. The team must be involved and contribute their expertise to each function. These include the effort and cost estimates, schedules and other commitments. This involvement is based on the existence of an SOW that includes the scope, technical goals and objectives, and identification of the customers or end users. The SOW should be developed from the WBS in accordance with the company's imposed standards, the assigned responsibilities based on capability, the cost and schedule constraints, and the identification of dependency on other organisations. Without exception the SOW must be approved by the company management and the respective leaders involved in this project, product or service.

* Based on a Vision of the Company's Future
  – A Blueprint for the Action & Results
* Everyone Should be Involved and Informed
  – Everyone Should Get Involved
* Based on Specific and Measurable Processes
  – Real Inputs and Outputs are Discussed
* Objectives at Every Level Support Overall Goals
  – Fulfilment of the Company's Goals

Chart 7.11   What a Plan Should Be. Note: See Appendix section for full page image

Again the importance of the mentoring, coaching and training responsibility role comes forth as a requirement of all good leaders. All personnel involved in estimating and planning activities should be fully trained according to the needs of the company. This includes all leaders, managers and functional personnel.

7.2.4   Subcontractor management

The real purpose of subcontractor management is to identify the best qualified to do the work required when the team has decided that it will depend on an outside source. The requirements and configuration must be clearly communicated. Agreement must be reached as to how the subcontractor will deal with changes to the baselines and how these will be accounted for by both the prime company and the subcontractor. A communication system must be developed and agreed to by both parties. Part of the contracting agreement must concern the tracking process that the prime company will use and its agreement by the subcontractor.

The subcontractor must understand what work package they are building for and how it will fit into the overall product when assembled. The same can be said of the overall work package. It is incumbent upon the team to share the operational definition and the risks identified with the subcontractor. Where there is a working WBS this should also be shared.

Subcontractors must be qualified to do the job and work required. It is incumbent upon leaders and managers to assess the qualifications of the employees to be assigned to this work package. The qualifications are of course the capabilities of the personnel on the subcontractor's team. The same requirements must be placed on the subcontractor's team as on the functional operations or manufacturing team of the prime. All support organisations should be subject to tracking and oversight of performance results.

The author recommends that the prime company establish a system of contracting subcontractors in which a documented agreement covering both technical and non-technical requirements is established and clear to all involved. The prime company must ensure that all the functions that pertain to them is understood and carried out by the subcontractor. Items such as planning, tracking and oversight, configuration, quality and adherence to requirements should be treated with the same capability as by the prime company itself.

The leader of the project should be assured by documented proof that the subcontract manager is knowledgeable and experienced in the fields required, as well as supported by capable individuals assigned to the work package. There should be an understanding that the subcontract manager will coordinate and be responsible for the terms and conditions as drawn out in the agreement between the two parties. In addition, where it is known that some of the individuals assigned are not fully knowledgeable, there should be a mentoring, coaching

1. Create the Operational Definition for the Company
   - What is the Product, the Vision & the Mission?
2. Develop a Risk Management Strategy
3. Build a Useable Work Breakdown Structure
4. IDs for the Task Relationships
5. Estimate the Work Packages (Time & Cost)
6. Calculate the Initial Schedules
7. Assign and Level the Available Resources

Chart 7.12   First Steps of a Real Plan. Note: See Appendix section for full page image

or education programme planned, at the subcontractor's expense, to bring them up to speed (Chart 7.12).

Where there is a lack of knowledge about the relationship between the prime company and the subcontractor, a training orientation programme should be set up for all of the affected parties. This meeting of minds and personalities will do a great deal to establish a better relationship and more positive output between both groups. There needs to be a general understanding of prime company's standards and normal expectations when working with another party. This can best be accomplished through an orientation session attended by both organisations.

7.2.5   *Quality assurance*

Quality assurance gives the leader an appropriate mechanism for review and audit of the work packages from both the prime company and the subcontractor, to verify whether they comply with the applicable standards and procedures of the project. Measuring these qualities ensures that the items will fit into the overall work package as designed by the team, especially where there is a definitive life cycle determined.

Now we have to look at the hard part. Someone must play 'hardball'. This is either the management or an organisation established by management to look at the quality issue and establish that it is being done. This goes further than simple inspection. Quality assurance (QA) has to verify that standards are being followed, processes are being used and that only a very small amount of output is being rejected from the production floor. These people are the eyes and ears of the leaders and senior management, doing performance appraisals and reviewing the steps being taken to bring the product to fruition. These individuals will also have the confidence of senior management when they are doing their job.

This type of role requires a good deal of training. It is therefore incumbent upon the leaders and managers of the company to ensure that each member of the QA team is well trained, and that they understand their responsibilities, the methods that are used and the tools that they can use in their assessments. Management must define the capabilities and give top priority to interpersonal communications as a skill required of the personnel in the QA group.

The most common issue brought before the technical or non-technical performer on the work package is this: 'What process are you using and how are you applying it?' If the performer is truly doing their job the answer should be simple. It is when the performer does not know what process they are following that the project gets into trouble. Again, the leader should be aware of the operations and help the performer execute their role through mentoring, training or education. The best result is most often mentoring.

Assessments by the QA group will most often provide a strong 'well done' or a corrective action report. It is important for the leader to be receptive to the corrective action reports and to take steps to correct what they have found. These actions are necessary, and they help the stakeholder to improve, often by increasing their personal capability on the job.

7.2.6 Project tracking and oversight

The purpose of project tracking and oversight is to give adequate visibility to the actual progress of the work package. This allows the team to act effectively when progress has deviated from the actual plan. It is the duty of the stakeholders to compare the actual results and performance with the plan, looking for variation or deviation as they progress. This allows for corrective action and changes to the product commitments when agreed to by all the members of the team. Remember that the QA team is part of this overall team. If the corrections affect the requirements or configuration, then the CCB should be consulted for assessment and acceptance. Where this has occurred all the affected personnel and teams should be informed of the changes and adjust accordingly.

It must be emphasised that this data is based on the actual results and performance against the established plans. The feedback data is the responsibility of the project leader and will go directly to them in the event of any variation or change to the plan. It will then be the leader's responsibility to take corrective action based on that feedback. If the corrective action requires the violating individual to be disciplined or the process or plan to be changed this will be the leader's responsibility. A process or plan change will require the involvement of the CCB and senior management; the leader's next step will therefore be determined by the assessment of the type of corrective action required (Chart 7.13).

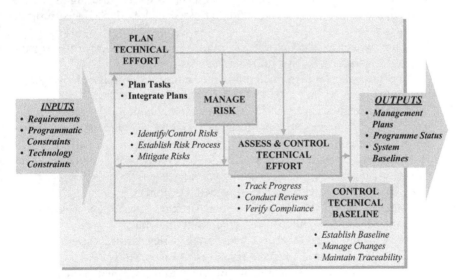

Chart 7.13   A Systems Engineering Management Process. Note: See Appendix section for full page image

Following each corrective action and any changes approved by the CCB, all affected individuals and groups must be informed of the changes made and approved. Without this information, many operations may continue to operate as though there are no changes, and so make faulty product. The information provided to the affected groups and individuals helps the company improve and produce better products. Most of these actions can not take place if there is no QA group to monitor and report on the issues that are occurring. The author encourages leaders and companies to establish such a group, or at least a group of persons who might have these responsibilities for the good of the product, project or service.

As in previous chapters, it should be pointed out that the leader is not restricted to the use of the key processes in level two. Two processes are strongly suggested from level three: training and peer review. Use of these two processes will make the operation of level two key processes so much easier, and will support their application. It cannot be said too strongly how important mentoring, teaching and coaching are to the successful development of the stakeholders in the company. These are skills that can best be developed as the company develops. Peer review is also a learning process that gives the stakeholder a means learn lessons from others as the project, product or service progresses [5].

Questions for the reader

1. Are requirements baselined in your organisation and under configuration control?
2. As requirements change, are adjustments made to the engineering plans, the configuration where necessary, and other associated work products (project, product or service)?
3. Does the project, product or service follow the written policy of the organisation for managing the system requirements?
4. Are measurements taken to determine the status of the proposed changes to the requirements?
5. Does your organisation have a documented configuration management plan?
6. Are all the engineering work products, deliverables and work packages under some form of configuration control?
7. Does the project, product or service follow a documented procedure for controlling changes to the engineering work packages?
8. Does your company have a CCB or something similar?
9. Are standard reports on the engineering baselines and change requests circulated by the CCB to all affected groups and individuals?
10. Is there a documented QA plan in your company that assesses the engineering processes, and are there adequate resources to support this?
11. Are the results of reviews and audits provided to the affected groups?
12. Is there a process in your company for determining the qualifications for subcontractors, and are the results of reviews and audits shared with them?
13. Is there an agreed-to policy and procedure for changes to the baselines and configurations that means that the subcontractors receive notice and information critical to their work?

14. Is there a procedure in your company that allows for or requires the work packages' actual size, schedule and cost to be compared to the estimates?
15. What form of corrective action is taken when performance results, open issues, unforeseen risks and action items are brought to the fore?
16. Does the work package have a technical management plan or engineering development plan that includes expected cost, size and schedule?
17. Are measurements used to determine the status of activities identified in the work package plan?
18. Does your company have a storage or library facility that can be used for the sole purpose of configuration management? Can you explain how it works?
19. If you do not have such a storage facility, how does your company control its requirements configuration?
20. Can you determine the difference between a project, product, service, work package or deliverable?
21. What role does the statement of work have in the overall project configuration? What role does it play at your company?
22. What role does QA play in your company's operations?
23. Does your company have a tracking and oversight process in place for its interactions with subcontractors?
24. How are work packages determined in your company?
25. Are responsibilities assigned to each work package in your company?

References

1. Guderian, B. *Leadership Development and the Role of Continuing Education.* Presented to graduating class of the ELITE Program; Tulsa, OK, 2010, p. 2
2. Humphrey, W.S. *Characterizing the Software Process: A Maturity Framework.* Software Engineering Institute, June 1987, pp. 1–20
3. Verzuh, E. *The Fast Forward MBA.* 2nd edn. New York: John Wiley and Sons; 2005, p. 115
4. Institute of Configuration Management. *CMMII Versus Other CM Certification Programs* [online]. 1988–2004. Available from http://www.icmhq.com/cmii-whitepapers.html [Accessed 6 May 2013]
5. Software Productivity Consortium. *Capability Maturity Model, Version 1.1,* Carnegie Mellon University, 1991, pp. 2–1 to 2–25

Chapter 8

Recommendations for process and capability in today's industries

Figure 8.1   P-6 ready for flight (1934)

Using a process associated plan and the concept of capability provides skills that are important to leadership. The development of the ELITE Leadership Model arose from these concepts and fundamentals. The personal capabilities and operational processes are repeated when they are guided by a user or leader who understands the concepts. This is done with more definition and useful techniques to make it easier for any future leader to develop their skills in a more appropriate way. In this chapter the author provides recommendations on the appropriate use and tactics of good capability skills and process applications (Chart 8.1).

8.1   Recommendation 1

It is important for an aspiring leader to study each of the four skills that make up the ELITE Leadership Model and attempt to develop their own personal skills in all four categories required to apply them. This can only be done by carefully studying the model and applying the skills on the job in situations that one encounters, while considering carefully the co-workers and stakeholders who are working with them.

The ELITE Leadership Model is made up of four distinct categories or skills:

- The self and the skills required for knowing oneself. These include self-awareness, social awareness, personal relationships and self-management.
- People skills include the development of people, establishing effective teams, having functional courage and motivating others.
- Operational skills include the management of key processes, business acumen, project and systems applications, and effective business judgment in daily operations.
- Organisational skills include the vision, strategy and mission, an enterprising perspective, change leadership skills and the appropriate organisational alignment.

The self requires a personal awareness, a social awareness of how you react and appear to others, the positive relationships that are developed and an ability to manage them in a productive way, which are all important. People skills require the courage to operate in the manner one knows will work especially well with others (this is developed from experience), the ability to develop others in a positive manner that will benefit both the company and the stakeholders themselves, and the ability to motivate others, especially in developing meaningful and successful teams. This includes the idea that leaders should be mentoring, teaching and educating their followers as involved stakeholders.

The operational skills are those which are the most overlooked and least often used correctly by engineering leaders. They include a productive understanding and

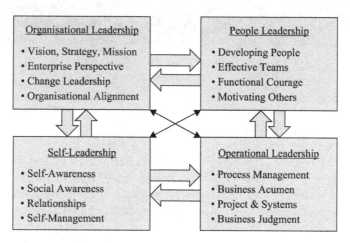

Chart 8.1   The ELITE Leadership Model [Source: Reprinted with permission of the University of Tulsa, ELITE Program]. Note: See Appendix section for full page image

management of the key processes, including: requirements, configuration, subcontracting, quality assurance, planning and tracking and oversight of the plan. Business acumen and application of systems and process engineering are included. These aid the appropriate business judgment for the company. Business judgment does not come easily to any leader; it is best developed over time through carefully developed experiential functions that are assigned to the potential leader as they develop (Chart 8.1).

The last part is understanding and applying the organisational skills. These include the enterprise perspective, a development of the company's application of the vision–mission–strategy, and awareness of the desired organisational alignment. The ability to deal with change and function in a changing environment is a prime concern. Today's leaders cannot function effectively in the business environment without understanding the importance and function of change management. It is important to keep up with the studies being done in industry, especially today. Conlon, Isler and Kochanski have shown in a recent study [8] that organisational affiliation is a very important component in the retention of staff when a company has grown large enough to be of concern to the stakeholders. This study shows the importance of applying personal development for stakeholders as well. If a leader truly demonstrates the skills required to be a leader, they will be looking at how to apply the mentoring, coaching and teaching skills to the people that work for them, and be emphasising the identification of variation, customer concern and supplier issues as items that both new and experienced personnel need to develop to be productive players in the company. The delegation and responsibility of requirements will thus be done in a functional and facilitating way that ensures that the skills required by the company will be applied and that they are being learned and applied in a progressive manner.

Today, employees and stakeholders will not allow themselves to be continually used to better the company if they themselves feel there is no opportunity to grow as well. This reality is often overlooked by the leadership. The stakeholders' solution is to leave and find other jobs that provide what they are looking for. If you study the culture differences researched and supported by Marilyn Moats Kennedy, you will find that the results point to the fact that today's generations choose to leave and look for what they consider to be 'greener pastures' [7]. Mentoring, coaching and teaching will often constitute the development and continuing education of the employee that they most often want or need. If this opportunity is not available, the company will notice a reduction in force that is not by their own doing, but of those who are leaving the company for other locations that do provide those opportunities.

8.2   Recommendation 2

Personnel being developed by the company should demonstrate an appreciation for the guidance given by the experts who are mentoring, teaching and coaching them. People who work for others develop a great respect and appreciation for those who try to lighten the workers' load and help develop their skills in a way that supports the leader. Some of the best advice is provided by William Lareau in *The American Samurai* [1]. He advises that the use of earned respect and approval to motivate is most appreciated by the leader who develops others to follow, and therein guides the individual or group to achieve the company's objectives. Lareau says that employees who are treated with respect and dignity will perform best, and in addition will believe they can achieve the goals set before them [1]. Leaders who are out there doing, teaching, coaching and living the accepted processes are demonstrating a consistent pattern in their application to the needs of the company and the expected behaviour that supports that, and they are demonstrating the appropriate processes and procedures to accomplish what needs to be done.

Lareau goes on to say that those who are taught to appreciate the skills of variation assessment are most often able to reduce cost for the company and demonstrate an appreciation for what the company is attempting to do. This affects the customer, supplier and the overall process. He goes on to say that the leader should be assessing their employees' work in real time, helping them to improve while the work is being done, when guidance is most useful. It should not wait until the performance appraisal time, which is the process used by many companies [1]. Real-time assessment speaks to the needs of the stakeholder and employee with a louder voice than the one-time performance review approach.

If you have focused on the employee developing the appropriate skills and doing the expected job, the free time that results will allow the leader to develop their own skills and knowledge by reading about and studying the current trends so that innovation and new approaches can be developed as they are visualised. Without this free time from the rigours of management, a leader is bound by the work at hand as opposed to thinking about what can be done to improve the product and the service that is provided. Amazing as it may sound, when an employee really understands the issue of variation, they are often the ones that recommend changes for the processes to the leader. The stakeholder is so much closer to the process and sees all of the functions as a fully applied system. If it can be improved, and they feel that the leader is receptive, they will make the suggestion. A good concept and tool to keep in hand is that taught by the late William Oncken. In *Managing Management Time*, Oncken uses the 'monkey analogy' to support the delegation of work to others and the use of management time to develop their own personal and working abilities to manage their personal time and work requirements in a meaningful way [2]. If we see ways to get the work done through others, and they are capable of carrying the load required (the monkeys), the work will get done and done well. Making sure that others carry that load (the monkeys) is the lesson that Oncken wanted all managers and leaders to learn.

The concept of respect and doing one's best to develop others is a process that is also encouraged by John C. Maxwell in *The 21 Irrefutable Leadership Skills* [3]. He and other leadership experts point out that developing others while in a leadership position will encourage respect and loyalty from that participant while developing their personal ability to perform the job [3]. When an employee feels that they are being developed based on discussions that they have had with the group's leader, they will be strongly loyal and hold that person in high respect because of what they feel has been done for them. However, this is but one item that spurs respect and loyalty. Others include the fact that the leader 'walks the talk' and really does what they say they are going to do. In addition there is the leader's example in standing up for what is considered to be right for the group and the organisation.

Maxwell sees this as a collection of skills and abilities, which he calls 'the law of process'. He sees this as a grand process and encourages leaders to follow it. His law encourages the development of those who work for you [3] to be constantly on your guard to find ways that the individual wants to develop, and to focus on those skills that will make them a better employee in their role. This tends to mature the people or stakeholders that work for you. Maxwell also points out that a true leader has to make this a cultural issue in which all those who report directly can expect to have this opportunity when they work for you [3]. Without question this type of change will affect all the individuals in your group once it is established. This is by no means an easy task to accomplish. Often the very culture that you are trying to change will be fighting you at every turn. But in the long run the effort is worth it. Without question, leaders invest in those who follow them, and the long term results are that those who do follow will return the favour ten times over in productivity and results [4].

- A Primary Focus Must be on Teaching & Coaching to:
  - Control Processes & Keep Work Distributed
  - Satisfy the Customer
- Develop Long-Range Customer Driven Plans
  - Involving all Employee Inputs
- Devote Time to Studying, Thinking & Learning
  - About Bold New Innovations
  - Concern for the Success of the Company

Chart 8.2   The Prime Management Function is Leadership. Note: See Appendix section for full page image

When leadership is the primary function, individuals look at appropriate means to develop others and themselves. One important factor is the ability to spend time reading, studying, learning and thinking about the types of changes that can be brought to the company in the form of innovations (Chart 8.2). There are many concerns around the field of engineering, but the most immediate ones are leadership and innovation. In 2006, the American Society for Engineering Education (ASEE) invited several company representatives to their annual meeting. This is held around the country, and that year it was in Honolulu, Hawaii. The company representatives came to present to the ASEE on the theme of 'Industry speaks with one voice'. The most common concern presented during that presentation by the companies there – Northrop Grumman, Boeing, Lockheed Martin and Raytheon – were that they found most of the graduates from the engineering universities to be lacking in two skills: leadership and innovation [11].

This inability to demonstrate these skills may be partly due to the fact that most universities do not focus on developing these attributes. But to examine the requirements of an innovator we find that most of these qualities are part of the make up of an engineer. They have a large appetite for information and the ability to visualise and develop new opportunities (Chart 8.3).

* Problems Begin & Continue Because no-one Teaches the Employee How to Control or Master the Processes
* If the Employee Looks to the Manager to Solve their Problems they will not Understand:
  – Variability
  – Supplier – Customer Links
  – Process Improvement
* Teach, Coach & Guide on a Real-Time Basis
  – How are they Doing NOW, Instead of Once a Year

Chart 8.3   Teaching & Coaching. Note: See Appendix section for full page image

An article published in *The Futurist* in 1987 spoke of the skills one could use to be a true innovator. These were the observational skills that enable you to draw your own conclusions on how things can be done and to ask questions to resolve unknown issues [9]. Be a trend watcher and make your reading time count. Watch the media for ideas and spot opportunities for change.

The rules presented in this article were:

1. observe the trends and develop ways to exploit them,
2. search for solutions to negative trends,
3. look at your activities, beliefs and interest for ideas to appeal to others,
4. come up with new ideas that run counter to current trends and
5. watch what the competition is doing and do it better [9].

8.3   Recommendation 3

The developing leader is encouraged to read and develop an appreciation for the concept of the Capability Maturity Model Version 1.1 (CMM) discussed elsewhere in this book. The author sees the five stages of capability as fundamental to all of the engineering fields. The second level or stage of the model sets the foundation and the key concepts that make up the driving forces for effective engineering applications that support project and programme management. They are:

1. requirements management,
2. configuration management,
3. sub-contractor management,
4. quality assurance,
5. project and programme planning and
6. programme tracking and oversight.

All the key processes function together to make up an effective and repeatable project format for the leader [5].

The CMM has a lasting effect on those companies that embrace it and work to make its key processes function for them. As Northrop Grumman (NGC) has demonstrated in its applications of the model [6], the operations of the organisation improve and function well on their projects and programmes. Through NGC's efforts and the integrated engineering process, the smoke stack approach to operations has been replaced with a functional activity system of process-oriented organisations.

As demonstrated by Chart 8.4, NGC's intention was to replace all of the discipline-oriented stovepipe document requirements with the activity orientation of a capability model. This help them directionally find the key processes that would make them more efficient and effective in their programme operation.

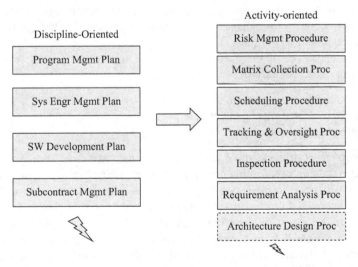

Chart 8.4   *Replaces Traditional Stovepipe Docs [Source: Reprinted with the permission of Northrop Grumman Corporation. From Freeman, D.G., Hinkey, M.E., Martak, J.W. 'Integrated Engineering Process Covering All Engineering Disciplines'. Presented at SEI Conference; Pittsburgh, PA, 2002]. Note: See Appendix section for full page image*

The next level beyond the repeatability level is level three: defined, or engineering process criteria. These include:

1. organisational process definition,
2. organisational process focus,
3. peer reviews,
4. training programmes,
5. inter-group co-ordination,
6. product engineering and
7. integrated management.

These key processes make up a series of important activities that allow an engineering organisation to become more defined and interactive with all of the parts of the overall company. When a company becomes more defined, it is able to know what activities are important. This definition also allows for planned peer reviews, a training programme that deals with the stakeholders and the inter-group coordination of all activities. Product engineering and integrated management are all coordinated as key processes with activities that support those functions.

The fourth level (managed) focuses on quality issues as a management or control function. These are known as

1. quality management and
2. quantitative process management activities.

The fourth level allows for greater control of the organisation in the area of quality and the ability to manage the system more effectively through the use of quality management techniques.

The highest level of the CMM (stage five – optimizing) is the continuous improvement level. Its activities include:

1. process change management,
2. technology change management and
3. defect prevention.

The key factor of the CMM as one moves up the chart from level two to level five is that risk is reduced on the projects, programmes and services for those who use the key processes in their everyday operations. The most useful processes the author has discovered are those at the second and third levels. These are directly applicable to the everyday functions that we deal with in engineering and the problem applications that we see on a day-to-day basis. Just the fact that these will reduce 50% of the risk makes them well worth the effort, and the return on investment is immeasurable.

8.4   Recommendation 4

There is a definite need to focus on the idea of developing the people who work for you. These are the stakeholders who can make or break your personal and company

Level		Focus	Key Process Areas	Result
Improve → **Optimising** 5		Continuous Improvement	Process Change Management Technology Change Management Defect Prevention	Productivity & Quality
Control → **Managed** 4		Product and Process Quality	Quality Management Quantitative Process Management	
Defined → **Defined** 3		Engineering Process	Organisation Process Focus Organisation Process Definition Peer Reviews Training Programme Inter-group Coordination Product Engineering Integrated Management	
Repeatable 2		Project Management	Requirements Management Project Planning Project Tracking & Oversight Subcontract Management Quality Assurance Configuration Management	R I S K
Initial 1		Heroes		

*Improvement initiatives **must increase market share** and/or profitability in order **to have business value!***

Version 1.1
C-SEI/CMU

Chart 8.5   A Process Management Model ... the Capability Maturity Model [Source: Image adapted with permission from Zubrow, D. 'Putting "M" in the Model: Measurement in CMMI'. The Software Engineering Institute Conference, Carnegie Mellon University], 2007. Note: See Appendix section for full page image

successes. Developing these individuals means adopting the concepts brought forward in *American Samurai* [1]. The author wants to establish an appreciation for measuring variation in such a way that allows for level two of the CMM and repeatability for the product, project or service commodity by the stakeholder. Without this form of personal development the stakeholders will not understand how variation measurement can be used to improve the processes being applied. Variation measurement reduces the potential for error and develops the individual's appreciation for effective and efficient operations. It should be noted that variation assessment is only a small part of the stakeholders' development needs in the overall appreciation of processes for requirements, configuration, subcontractor management and quality, as these bridge only a few of the basic processes. Understanding what a proper planning process is and applying tracking and oversight applications also top the list of process needs for the stakeholder.

The other more important part of development has to do with the ability to develop the body of knowledge for each job or role required in the specific group that you lead. When you know what is required to be successful and are able to determine the basic skill requirements for those roles, you also know what areas need improvement in the stakeholders' personal skills. This is often referred to as knowledge of the capability gaps. Knowledge of these gaps allows leaders to identify the kind of training, education and development required and to work out a development plan for those individuals. Over time, the development allows the stakeholder to grow and become a more valuable contributor to the team. It is important that the leader develops the skills to identify the body of knowledge' for each of the roles in their competency or capability grouping. These also allow the leader to become a highly efficient key or core competency provider within the company.

Here we are focusing on a leader's ability to take the identified core or key competencies, identify the specific competency that their group or team is best at, and to support and know what the body of knowledge is for each role within that team (see Chart 8.5). This can be considered 'real' leadership because it is the leader's knowledge alone of what makes the team great and supportive that will make the company better able to meet the needs of the product, project or service. Here too is where the ability to focus and appreciate change comes to bear. Resilience to deal with change, adapt what needs to be changed to the group and to meet the immediate needs of the company are what drive the true leader. As any facet of change hits this individual, he or she is able to adapt the group to meet the needs of the product, project or service.

Once we have developed an ability to know what needs to be done (understanding the competencies), and how it can best be done and with the most talented and appropriate personnel, we need to focus on the financial factors. How do we control the costs, and can we reduce costs through a change to the processes or procedures? Making these kinds of changes requires the loyalty of those stakeholders who you lead. How much have you developed your trust with your stakeholders? This is an indication of how well you have been 'walking the talk' and repeating the 'real' expectations while in their presence. Additionally, have you been doing your homework in identifying what really needs to be done? Have you been listening to your staff or the stakeholders in the field? Have they been willing to share with you their feelings on what needs to change? The real question here is: 'Have you been listening?' If your stakeholders feel they can trust you and speak their mind regarding any issues, have you been discussing their ideas with them and with others who have the controls of the processes in their hands? Once trust is in place the next step is to establish a communication process that lets stakeholders know that you have heard them and how these potential and realistic applications will be put into play. This is also where you need to know how and when the CCB needs to be in play and when not (Charts 8.6 and 8.7).

I am going to suggest a tool called the four panel status chart. This puts all of the features we have talked about into play. The four panel status chart allows the leader to place into the stakeholders' hands a playbook that says what their responsibility to the plan will be and the suggested timeline, cost and futures for them to react to and operate with. The four panel chart is a reporting mechanism that allows them to tell you what they are doing and how well. It also allows you as the leader to give them the necessary feedback.

The chart (see Chart 8.8) is available in the Appendix section of this book so that you can copy it and use it to communicate with the stakeholders in your group

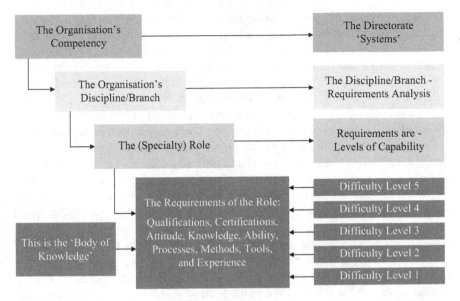

Chart 8.6 The Competency Structure. Note: See Appendix section for full page image

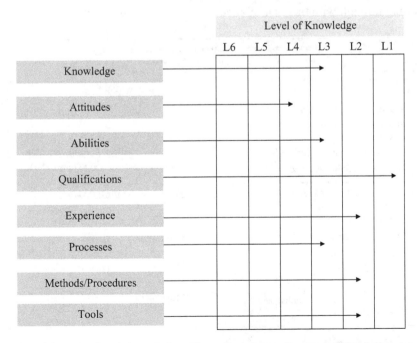

Chart 8.7 The Body of Knowledge. Note: See Appendix section for full page image

or team. Red, yellow and green colours are used to depict items that are in question (red), might be suspect and are not reaching the expected level of accomplishment (yellow), or are reaching the appropriate level of completion and accomplishment (green). Blue can also be used to depict completed milestones when it is important for them to be reported. When you want communication to take place, this is one of the best ways to do it. If there is a disagreement it will surface during the status meetings and give the concerned stakeholders a chance to react or comment. The four panel status chart is most often used by the management or leaders of the product groups, but it can easily be used in projects or for service operations too. The information that goes into these charts generally comes from the stakeholders in the form of a written report, or the use of the three panel status chart (Chart 8.9).

The three panel status chart has three very important components that allow the stakeholder to communicate with the leader. First, it allows the stakeholder to tell the leader what their six top objectives are for the product at the time of the report. It then allows the stakeholder to communicate the activities that are being achieved to the leader. Lastly it allows the communicator to tell the leader what the timelines are for each objective. If there is a problem or error in understanding by the stakeholder or the leader, it will arrive in the form of objectives that are unnecessary. Each objective has a colour chart that depicts the status of the objective and clearly shows where the stakeholder thinks they are with respect to the product, project or service they are listing. This form of communication allows the leader to query the stakeholder about their progress and to question whether the stakeholder is really following through with the plan of attack that was originally agreed.

A Four Panel Status Chart can be used to identify the development/execution of the capability plan	
A Accomplishments • List significant completions performed (eg. milestones, key reviews, critical hardware receipt, etc.) • List initiative gates accomplished • Critical cost points reached	**C Issues Watch List** Item Action Plan Closure Date • Any Critical Item w/trend of S • List is used to flag non-milestone items • Recovery plan required to return to O

B   Critical Milestones
Item   Date   Trend
• Status of Critical Stage gates
• Status of future Stages & Completions
• List important Initiatives
• Trend indicators are:
  • 0 = On track
  • S = Slip from scheduled date
  • A = Ahead of Schedule
	D Milestone Health Indicators
	
	                    Current +3Mos
	
	Milestone #1 Red Yellow
	
	Milestone #2 Yellow Green
	
	Milestone #3 Green Green

Chart 8.8   Four Panel Status Chart. Note: See Appendix section for full page image

The colour code used for is the same as that used in the four corner chart to reduce misunderstandings or confusion.

This chart is also available in the Appendix for the readers to use as a communication tool with the leader or management. Now the stakeholder has a mechanism with which to apply their successes, concerns and complaints in a written format that can be seen by the leaders and allows the two parties to communicate their differences or agreements on a very simple one-page illustration.

8.5 In conclusion

Now that all of the recommendations have been made, let's look back at each suggestion and remind the reader what they can do to be better leaders and more responsive managers in the mainstream of their company. It should be obvious by now that the most important factor is for the leader to be a coach, mentor and teacher to their people – the major stakeholders of the product, project or service they are overseeing. Once this is realised, the leader must understand how to apply the four categories or parts of the ELITE Leadership Model and attempt to follow the guidelines provided in this book. Operational skills might be considered unimportant and are often overlooked by leadership and management in most companies. However, they are necessary and need to be put into place on any project, product or service the leader has taken on.

Key: Blue – Task Complete, Green – Task on Schedule, Yellow – Behind Schedule but Doable, Red – Task in Jeopardy

Engineering - XYZ Corp

Objectives		*Activity*
1. Objective 1	Yellow	1. Activity 1
2. Objective 2	Blue	2. Activity 2
3. Objective 3	Green	3. Activity 3
4. Objective 4	Yellow	4. Activity 4
5. Objective 5	Green	5. Activity 5
6. Objective 6	Red	6. Activity 6

Schedule

Tasks	Jan	Feb	Mar	Apr	May	Jun	Jul	Aug	Sep	Oct	Nov	Dec
1. #1			△									
2. #2												
3. #3												
4. #4												
5. #5												
6. #6												

Chart 8.9   Three Panel Status Chart. Note: See Appendix section for full page image

Operational leadership leads to the CMM and the processes that are supported by this tool. Requirements, configuration, quality, planning, tracking and oversight and subcontractor management all provide the leader with the tools and skills for a more effective organisation. Inherent in those tools is the ability to develop the core or key competencies and to maintain them through the effective use of a competency structure in which the leader helps the company develop a body of knowledge for each job or role in the company. Once this is established the company can adjust the skill requirements for every stakeholder and establish the training requirements in a more productive manner for each person in the company. Knowing the body of knowledge required for each role allows the company to make its capability a sales issue with customers and the public. While some of what is being suggested exists on level three of the CMM, these are good concepts to be applied. In addition it is suggested that the peer review key process be implemented as this will allow for further feedback and input from the stakeholders throughout the project's progress.

The next factors have to do with getting the work done, controlling the change requirements and saving the company money in executing the processes and procedures. Once the stakeholders are involved and feel that their leader wants them to suggest changes as required, they will look for variation in the operations and make suggestions about where the correction best fits. Since the leader is 'walking the talk' the feedback is rapid and current, allowing the adjustments to be made and taken to the CCB for company acceptance. As the product, project or service sees a change in productivity and a reduction in cost, the company saves money, charges the customer less and everyone is happy. However, that is a perfect world and we don't live in that environment. So the leader must be on the lookout for any changes required and to have the loyalty of their stakeholders by thanking them for their input on a regular basis.

Change in a company does not come easily; it has a long history of resistance from the people involved in any operation. It extends from the lowest levels of the organisation to very highest. All have their reasons, and their realities are often the perceptions that they have developed over time from experience and lessons that they will claim taught them to operate as they currently do. That doesn't make them right or wrong; it only makes them a roadblock in the application of the changes that might make the organisation that much better in the long run.

Some of the resistance may come about because of the age of the stakeholder and the number of years that they have been with the company. Other types of resistances might be due to the number of roadblocks that they have experienced personally. As the leader have you asked them why they feel the way they do? Sometimes a simple discussion will provide a resolution to problems that have bogged down the progress of the project.

The key here is the resilience of the leader to listen, understand and make the necessary changes, and to stay vigilant to the objective and goals required. Saving the company money and reducing costs are all things that should be constantly on the minds of all the stakeholders, include the leaders, top management and employees.

Questions for the reader

1. How well have you been 'walking the talk' and repeating the expectations while in the stakeholders' presence?
2. Have you been doing your homework to identify what really needs to be done, such as the innovation and change to meet current needs?
3. Have you been listening to your staff or the stakeholders in the field? What are they saying? Does it make sense and can you give them the necessary feedback effectively?
4. What do you think you can offer that others are not? Does your innovation stem from some idea you developed, or from something that you saw others doing?
5. How do you currently add value to the product, project or service you currently work with?
6. What might your customers want that you are not now producing?
7. How might you produce your project or service more efficiently and at a lower cost to the customer?
8. Have the stakeholders been willing to share with you their feelings on what needs to change?
9. If you have really been listening, what are they saying about the processes? How have you been dealing with their suggestions?
10. What is the benefit of questioning the change required to do away with the company stovepipes?
11. How can CMM 1.1 be used to improve the processes in your company?
12. How will you use the competency structure to improve the requirements of your stakeholders' skills, abilities and body of knowledge?
13. Explain your understanding of the body of knowledge and how you use it in taking the capability requirements from the competency you represent. How do you apply it to the roles and jobs required in your group?
14. Does the Four Panel Status Chart make sense to you? How would you use it in your company?
15. How would you delegate the use of the Three Panel Status Chart to your stakeholders?
16. Can you explain how you would use the ELITE Leadership Model in your leadership role in company?
17. How are differences in understanding the objectives dealt with in your company?
18. Is there a set of rules available that allows the stakeholder to set certain objectives, or are they responsible for following the guidelines set by the leadership?
19. If a stakeholder spots a process or procedure that can improve the steps taken in production, are there rules in place that allow them to bring them forward in a status meeting?
20. How are the stakeholders allowed to present variance concerns to the leadership?

21.   Is there a real concern in your company for the use of requirements and configuration processes? If so, how are they handled?

22.   Have you, as a concerned leader, asked those resisting the necessary changes why they feel the way that they do? Is there some logic to their thinking that you need to research?

23.   Are subcontractors held to the same processes as the stakeholders in the company? If not, why is this so?

24.   How is planning handled in your company? Are there rules or guidelines set out for tracking and oversight? Are the subcontractors held to the same guidelines?

25.   How does the ELITE Leadership Model fit into the picture? Can you name the four parts of the model? Which one has been left out at your company? Do you know why, or is this reason just conjecture?

References

1.   Lareau, W. *American Samurai: A Warrior for the Coming Dark Ages of American Business.* New York: Warner Books; 1992, pp. 140–147

2.   Oncken, W. *Managing Management Time.* Englewood Cliffs, NJ: Prentice Hall; 1984, pp. 161–70

3.   Maxwell, J.C. *The 21 Irrefutable Laws of Leadership.* 10th edn. Nashville, TN: Thomas Nelson Publishing; 2007, pp. 141–52

4.   Maxwell, J.C. *The 21 Irrefutable Laws of Leadership.* 10th edn. Thomas Nelson Publishing; 2007, pp. 23–34

5.   Humphrey, W.S. *Characterizing the Software Process: A Maturity Framework.* Software Engineering Institute, June 1987, pp. 1–20

6.   Freeman, D.G., Hinkey, M.E., Martak, J.W. 'Integrated Engineering Process Covering All Engineering Disciplines'. Presented at SEI Conference; Pittsburgh, Penn., 2002

7.   Kennedy, M.M. 'Career Strategies'. Presented at ASEE College Industry Education Conference, ASEE, San Antonio, TX, 2007

8.   Conlon, R., Insler, D., Kochanski, J. *Rewards of Work Study.* Sibson Consulting; 2009. Available from http://www.sibson.com/publications-and-resources/surveys-studies/?id=252 [Accessed 6 May 2013]

9.   Waitley, D.E., Tucker, R.B. 'How to think like an Innovator'. *The Futurist.* May–June 1987, pp. 9–15

10.   Wysocki, B. 'Grooming Leaders', *National Business Employment Weekly.* 1981, pp.19–20

11.   Tidwell, J.P.'Industry Speaks with One Voice', Presented at ASEE Annual Conference; Honolulu, HI, 2006

Appendix

Maintaining effcctive engineering leadership: dependence on effective process

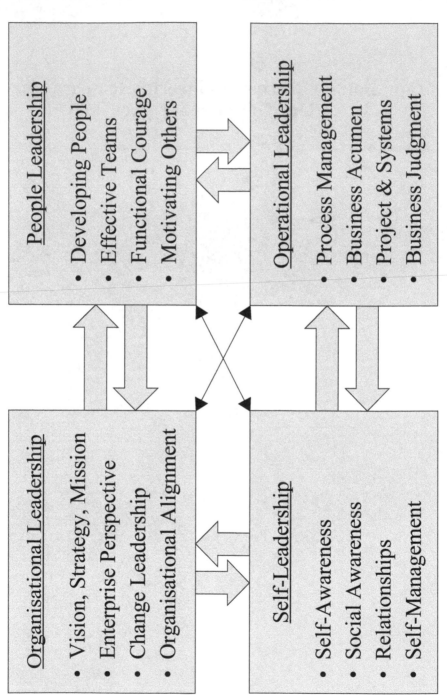

Chart 1   The ELITE Leadership Model [Source: Reprinted with permission of the University of Tulsa, ELITE Program]

Level		Focus	Key Process Areas	Result
Optimising	5	Continuous Improvement	Process Change Management Technology Change Management Defect Prevention	Productivity & Quality
Managed	4	Product and Process Quality	Quality Management Quantitative Process Management	
Defined	3	Engineering Process	Organisation Process Focus Organisation Process Definition Peer Reviews Training Programme Inter-group Coordination Product Engineering Integrated Management	R I S K
Repeatable	2	Project Management	Requirements Management Project Planning Project Tracking & Oversight Subcontract Management Quality Assurance Configuration Management	
Initial	1	Heroes		

Improve
Control
Defined

Version 1.1
C-SEI/CMU

Improvement initiatives must increase market share and/or profitability in order to have business value!

Chart 2   A Process Management Model . . . the Capability Maturity Model [Source: Image adapted with permission from Zubrow, D. 'Putting "M" in the Model: Measurement in CMMI'. Carnegie Mellon University, 2007]

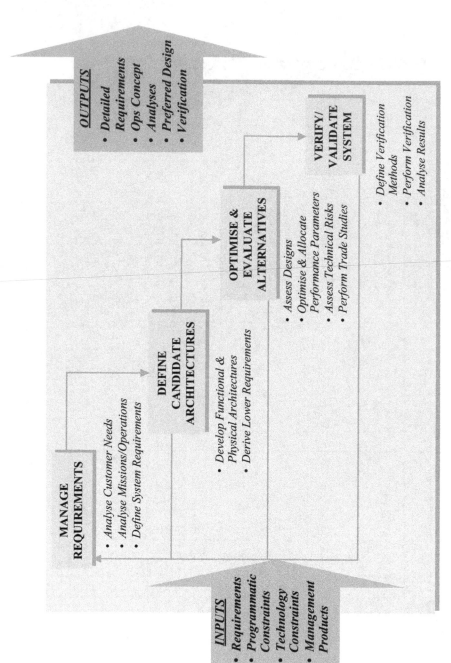

Chart 3   Technical Requirements Process

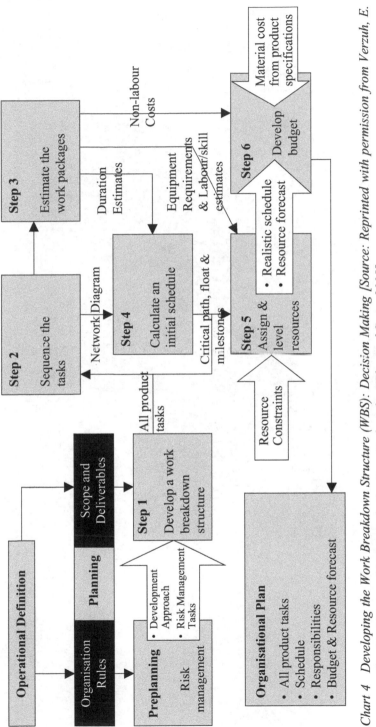

Chart 4   Developing the Work Breakdown Structure (WBS): Decision Making [Source: Reprinted with permission from Verzuh, E. The Fast Forward MBA. 2nd edn. New York: John Wiley and Sons; 2005]

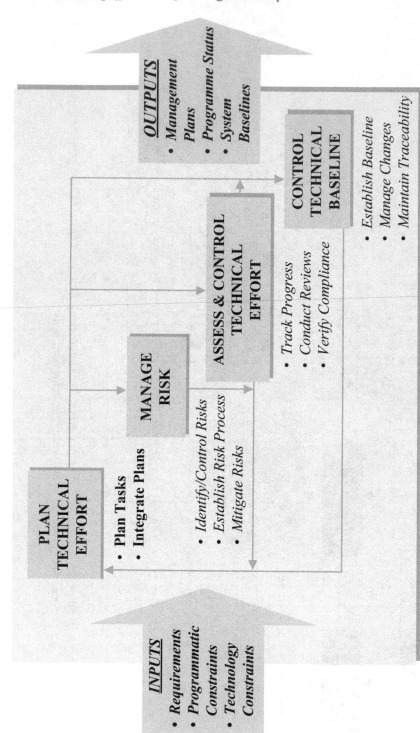

Chart 5   A Systems Management Process

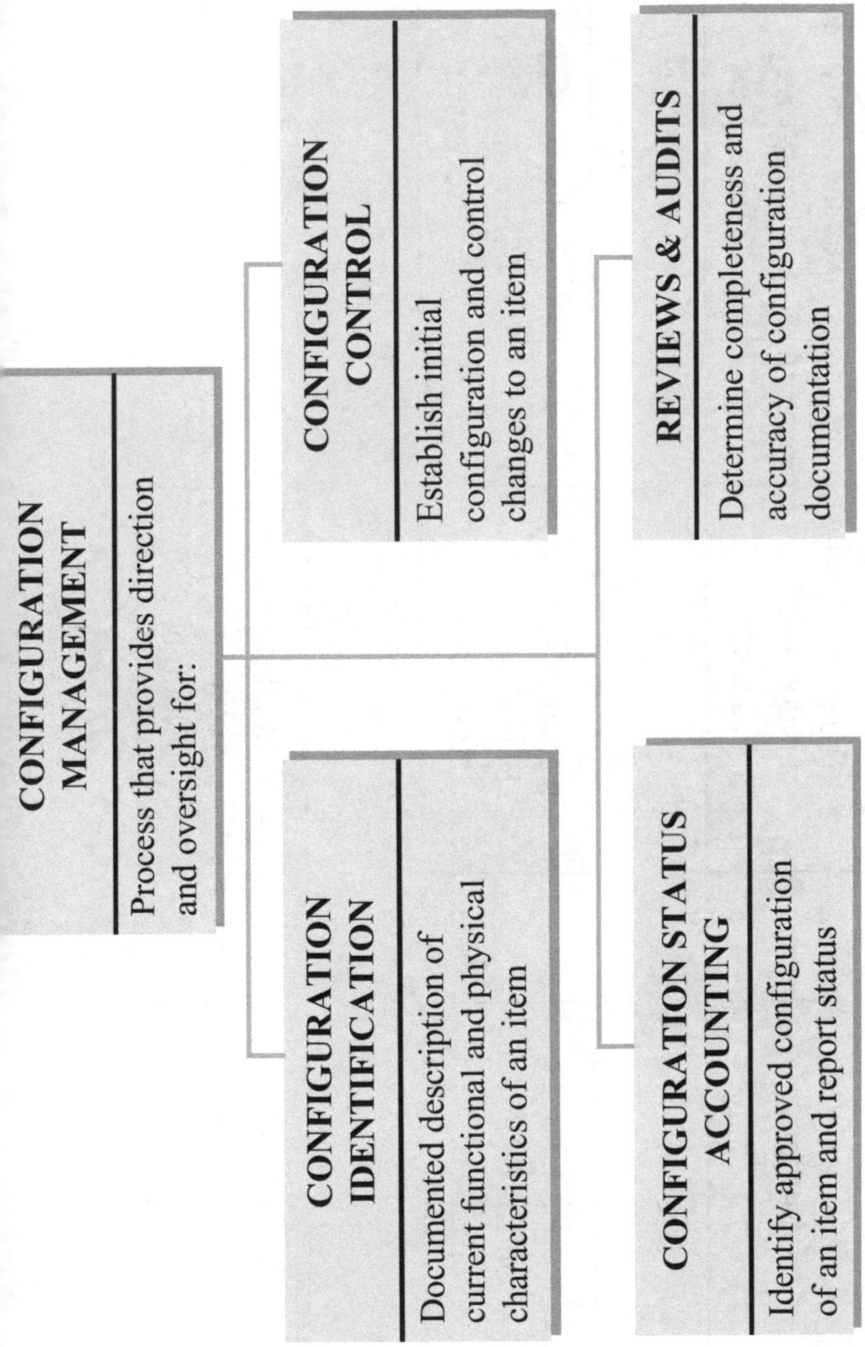

CONFIGURATION MANAGEMENT

Process that provides direction and oversight for:

CONFIGURATION CONTROL

Establish initial configuration and control changes to an item

REVIEWS & AUDITS

Determine completeness and accuracy of configuration documentation

CONFIGURATION IDENTIFICATION

Documented description of current functional and physical characteristics of an item

CONFIGURATION STATUS ACCOUNTING

Identify approved configuration of an item and report status

Chart 6   What is Configuration Management?

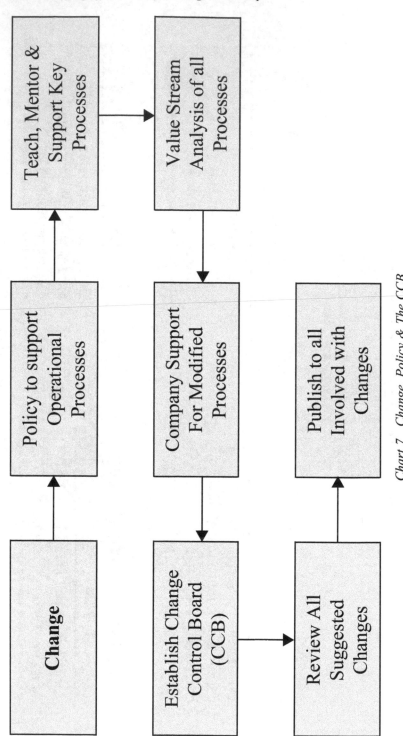

Chart 7   Change, Policy & The CCB

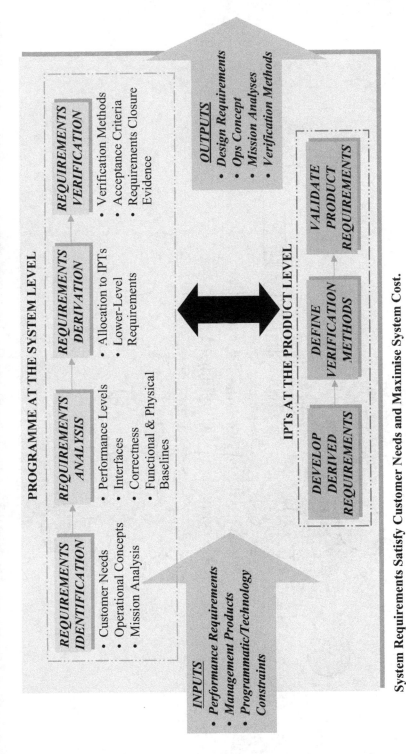

System Requirements Satisfy Customer Needs and Maximise System Cost.

Chart 8   Manage Requirements

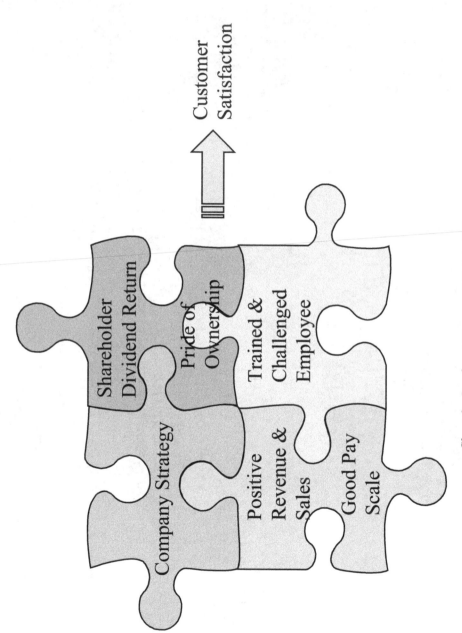

Chart 9   Leader – Managers' Point of View

- Based on a vision of the company's future
  - A blueprint for the action & results
- Everyone should be involved and informed
  - Everyone should get involved
- Based on specific and measurable processes
  - Real inputs and outputs are discussed
- Objectives at every level support overall goals
  - Fulfilment of the company's goals

Chart 10 What a Plan Should Be

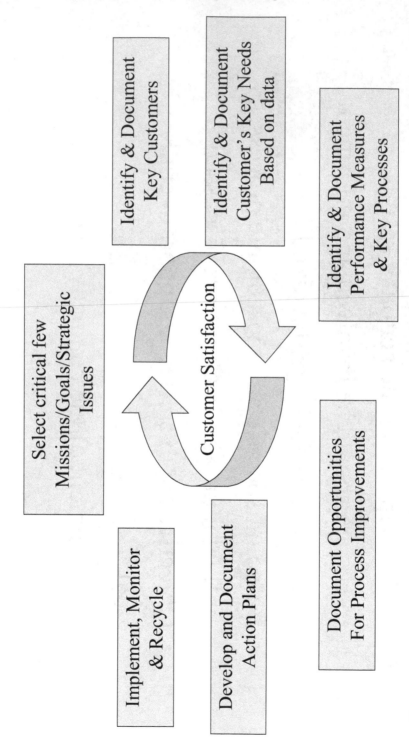

Chart 11   TQM Process Map (SAE Notes, December 1994) [Source: Copyright SAE International. Reprinted with permission]

- **Self-Awareness
  Development**
  – Self-Awareness Skills
    - Self-Awareness Skills
      of Effective People
  – Management Skills
  – Social Awareness
    Skills
  – Good Relationship
    Development Skills

- **People Leadership
  Development**
  – Effective Team
    Building
  – People Development
    - Coaching
    - Mentoring
    - Teaching
  – Motivation Skills
  – Functional Courage

Chart 12   Self-Awareness and People Leadership

- Problems begin & continue because no-one teaches the employee how to control or master the processes
- If the employee looks to the leader to solve their problems they will not understand:
  - Variability
  - Supplier & Customer Links
  - Process Improvement
- Teach, Coach & Mentor on a real-time basis
  - How are they doing NOW, instead of once a year

Chart 13   Mentor, Teach & Coach

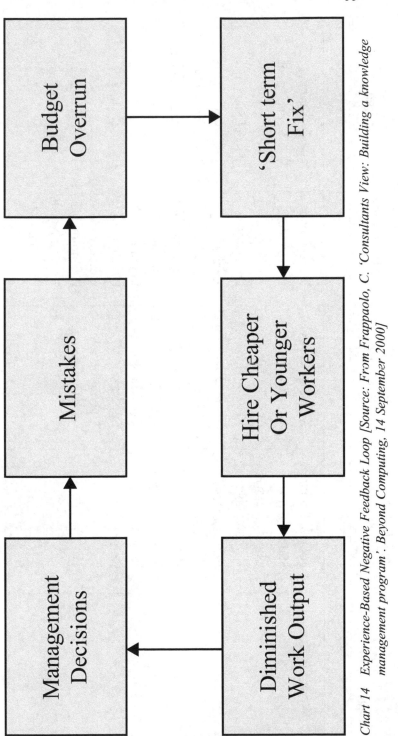

Chart 14 Experience-Based Negative Feedback Loop [Source: From Frappaolo, C. 'Consultants View: Building a knowledge management program'. Beyond Computing, 14 September 2000]

Characteristics Also known as: Born Between:	Pre-Boomers, Veterans, Silent Generation, Seniors (1922–1943)	Boomers, Me Generation, Sandwich Gen (1943–1960)	Generation X, Busters, Cuspers (1960–1980)	Millenials, Echos, Nexters, Generation Y (1980–2000)
Work Habits:	- Follow Tradition - Status Quo - Obedience over Individualism - Advancement Through Hierarchy - Sense of Duty & Honour - Natural Leaders	- Value of Personal Growth - Wants to be Involved - Team Orientation - Value Company Commitment & Loyalty - Sacrifice for Success - Uncomfortable with Conflict	- Entrepreneurial - Independent - Thrives on Diversity - Desires High Level of Responsibility - Constantly Looking for Creative outlets - Quickly Moves on if Employer Fails to meet Needs - Impatient	- 24/Seven - Capacity for Multitasking - Global Connections - Competitive - Civic Minded - Diverse - Desire for Structure - Goal & Achievement Orientation

Chart 15   Culture Differences [Source: From Kennedy, M.M. 'Career Strategies'. Presented at ASEE College Industry Education Conference, ASEE, 2007. Also available at www.moatskennedy.com]

- Tasks are not easy
  - Good chance for completion
- Team location is ideal
  - Reduced interruptions
  - Good facilities
- Tasks have clear goals to succeed
- Immediate feedback is provided
  - Adjustments are accommodated

Chart 16   Successful Teams [Source: From Csikszentmihalyi, M. Flow: The Psychology of Optimal Experience. New York: Harper Perennial; 1990]

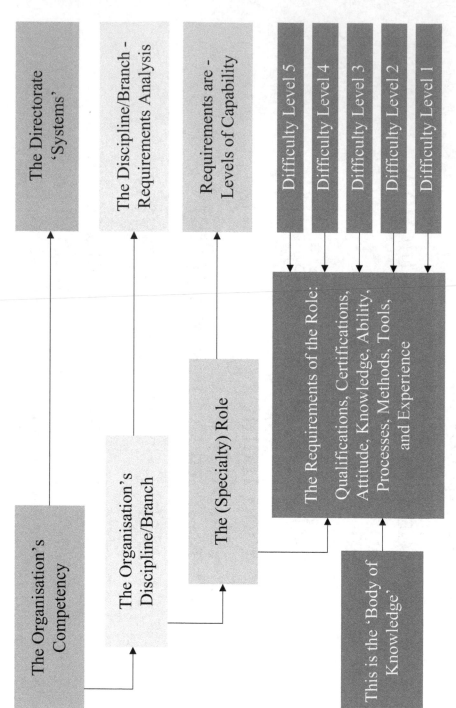

Chart 17   The Competency Structure

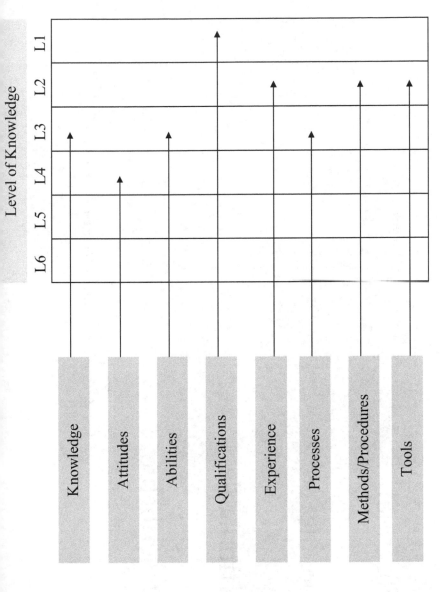

Chart 18 The Body of Knowledge

- Essential

- Doable

- Affordable

- Describable

Chart 19   Project Requirements Management Essentials

- Early Identification of Problems/Risks

- Work Around Solutions

- Cost Avoidance

- Schedule Slip Avoidance

- Improved & Effective Planning

- Application of a Professional Way of Doing Business

Chart 20 Proactive Approach to Process Leadership

1. Tolerated Innovation & Risk Taking

2. Attention to Detail & Reward System

3. Outcome Orientation

4. Impact of Leadership Decisions

5. Team Orientation

6. Company Atmosphere

7. Long-Term Perspective

Chart 21   Seven Characteristics of Organisational Culture

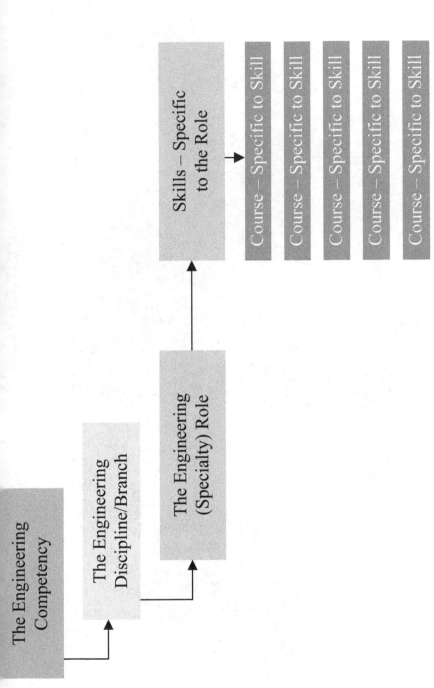

Chart 22 Simple Development of the Body of Knowledge

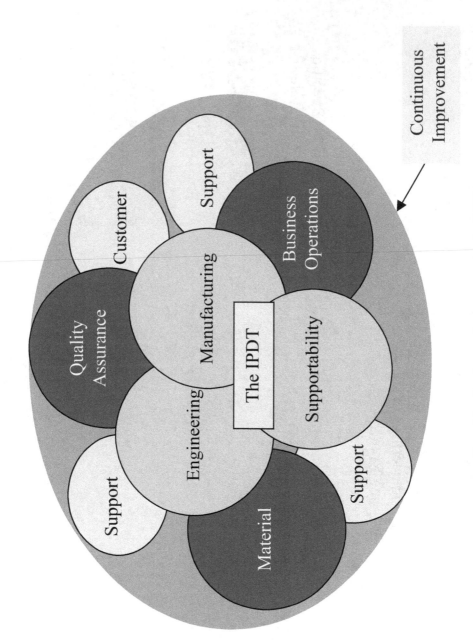

Chart 23   Continuous Improvement in the IPDT

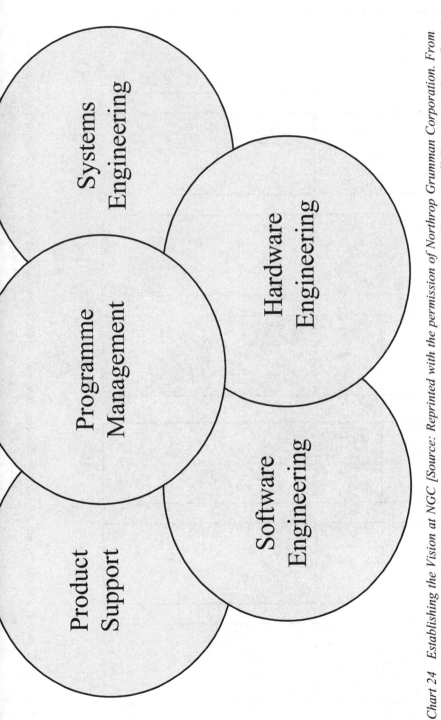

Chart 24   Establishing the Vision at NGC [Source: Reprinted with the permission of Northrop Grumman Corporation. From Freeman, D.G., Hinkey, M.E., Martak, J.W. 'Integrated Engineering Process Covering All Engineering Disciplines'. Presented at SEI Conference; Pittsburgh, PA, 2002]

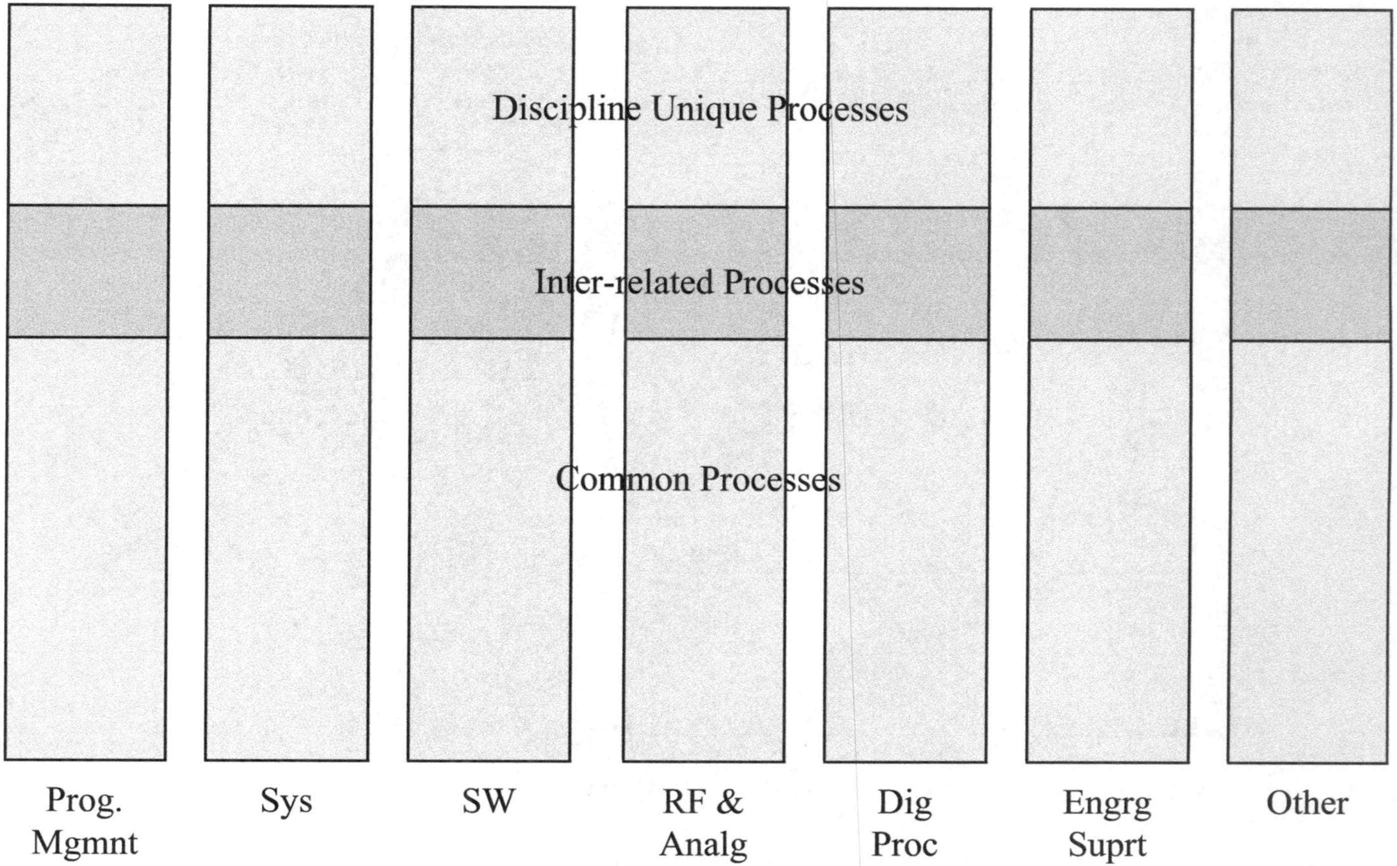

Chart 25   Common and Interrelated Processes [Source: Reprinted with the permission of Northrop Grumman Corporation. From Freeman, D.G., Hinkey, M.E., Martak, J.W. 'Integrated Engineering Process Covering All Engineering Disciplines'. Presented at SEI Conference, Pittsburgh, PA, 2003]

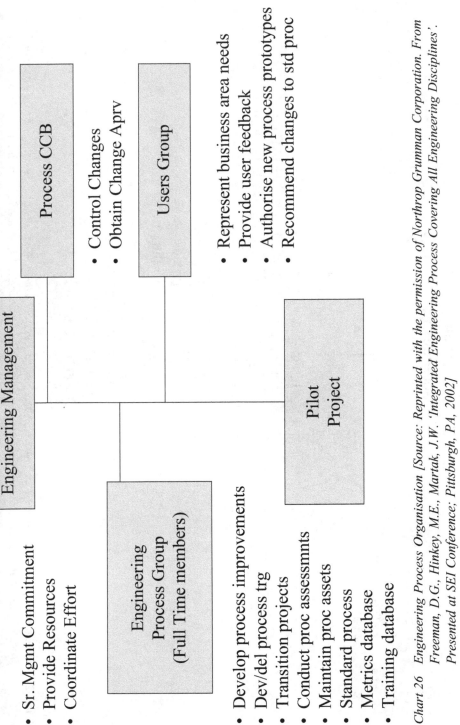

Chart 26 Engineering Process Organisation [Source: Reprinted with the permission of Northrop Grumman Corporation. From Freeman, D.G., Hinkey, M.E., Martak, J.W. 'Integrated Engineering Process Covering All Engineering Disciplines'. Presented at SEI Conference; Pittsburgh, PA, 2002]

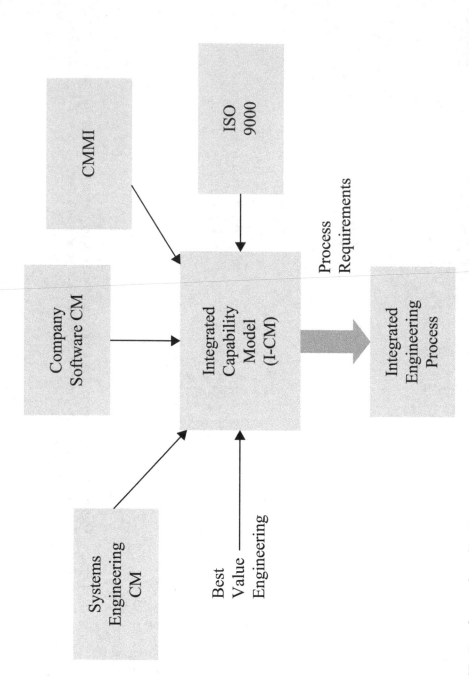

Chart 27   Developing the I-CM [Source: Reprinted with the permission of Northrop Grumman Corporation. From Freeman, D.G., Hinkey, M.E., Martak, J.W. 'Integrated Engineering Process Covering All Engineering Disciplines'. Presented at SEI Conference. Pittsburgh. PA. 2002]

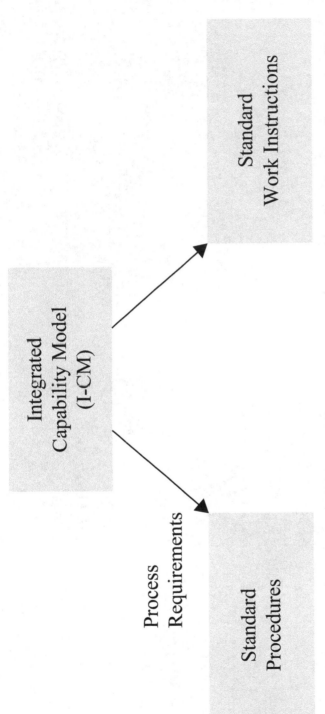

Chart 28 The Integrated Engineering Process [Source: Reprinted with the permission of Northrop Grumman Corporation. From Freeman, D.G., Hinkey, M.E., Martak, J.W. 'Integrated Engineering Process Covering All Engineering Disciplines'. Presented at SEI Conference; Pittsburgh, PA, 2002]

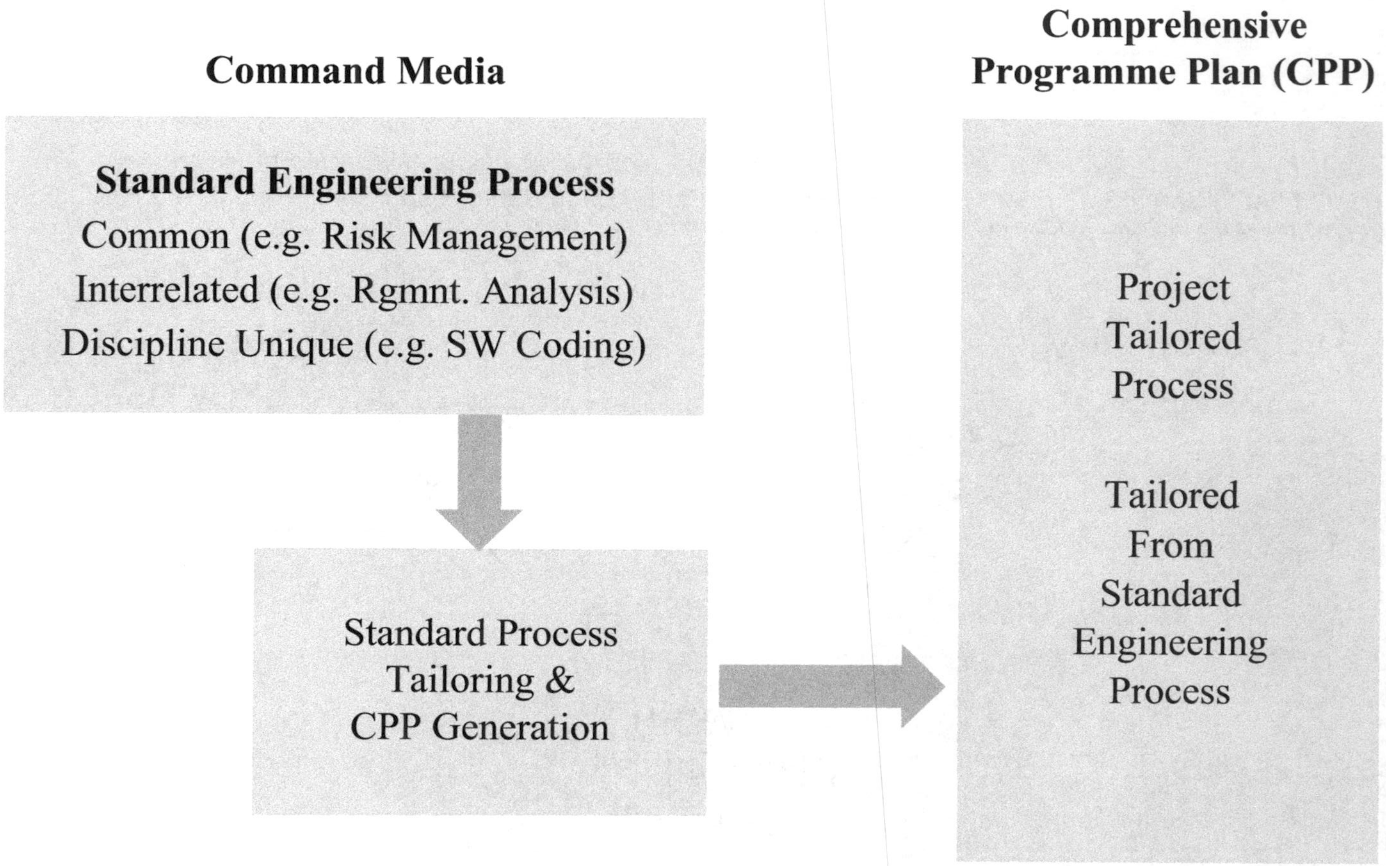

Command Media

Comprehensive Programme Plan (CPP)

Standard Engineering Process

Common (e.g. Risk Management)
Interrelated (e.g. Rgmnt. Analysis)
Discipline Unique (e.g. SW Coding)

Standard Process
Tailoring &
CPP Generation

Project
Tailored
Process

Tailored
From
Standard
Engineering
Process

Chart 29   Command Media to Programme Plan [Source: Reprinted with the permission of Northrop Grumman Corporation. From Freeman, D.G., Hinkey, M.E., Martak, J.W. 'Integrated Engineering Process Covering All Engineering Disciplines'. Presented at SEI Conference: Pittsburgh, PA, 2002]

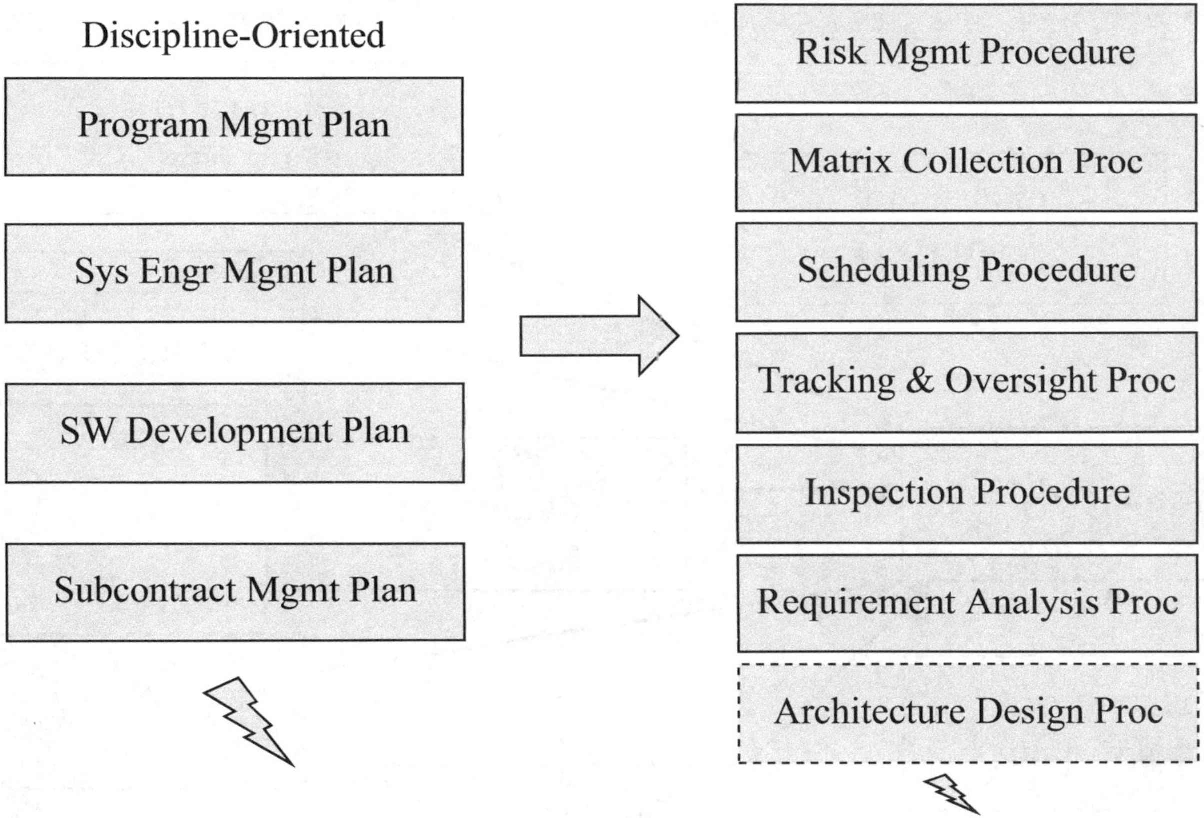

Chart 30   *Replace Traditional Stovepipe Docs [Source: Reprinted with the permission of Northrop Grumman Corporation. From Freeman, D.G., Hinkey, M.E., Martak, J.W. 'Integrated Engineering Process Covering All Engineering Disciplines'. Presented at SEI Conference; Pittsburgh, PA, 2002]*

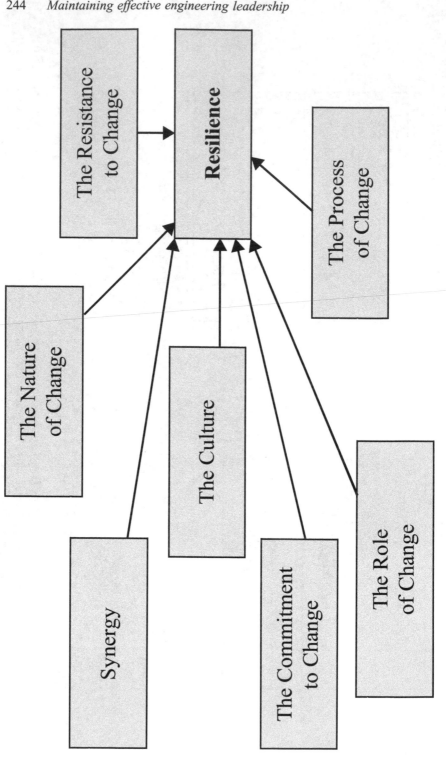

Chart 31   D.R. Conner Says That the Elements of Change are Best Met with Resilience [Source: From Conner, D.R. Managing at the Speed of Change, New York: Villard Books; 1994]

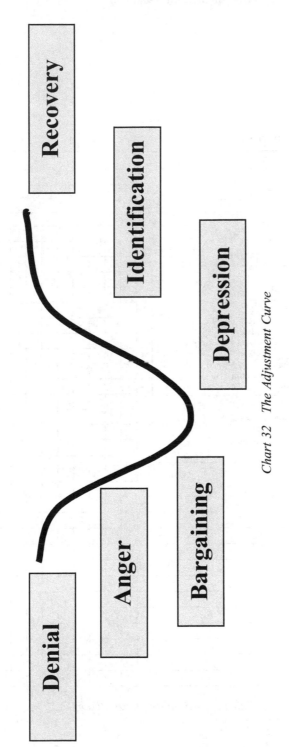

Chart 32 The Adjustment Curve

MONTHLY STATUS REPORT   AS OF: ___/___/___

Project Title: _____   _____

Principal Investigator: _____

_____   Prog. Code: _____

Cost Legend:

    Plan   = - - - - }
    Actual = ———————— } ————————→

Milestone Legend:
    ● Planned project start
    ◖ Subcontract initiated
    ◗ Subcontract completed
    △ Key analysis/model completed
    ◠ Data gathering completed
    ▽ Begin fact finding - Data gathering
    ✳ Major report
    ▽ Review/seminar
    ◉ Planned project work complete
    ◻ Final report due

Status Legend:
    ◯ Fill in milestone symbol if
        completed on schedule
        If missed, indicate slippage/
        new target by ◯⟶•

SK

Month →	O	N	D	J	F	M	A	M	J	J	A	S
Milestones:												
1												
2												
3												
4												
5												
6												
7												
8												
9												
10												
11												
12												

Chart 33   Monthly Status Report

- Select the Process to Benchmark
- Determine the Project's Scope
- Choose Relevant Measurements
- Study Performance Boosting Best Practices
- Judge Appropriateness & Adopt Practices
- Identify Cultural Issues/Other Implementation Factors
- Plan & Implement Changes
- Measure Results & Analyse Benefits

Chart 34   The Benchmarking Process

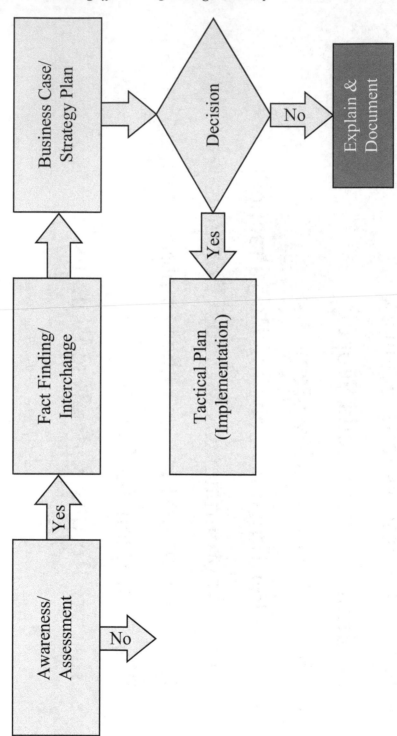

Chart 35   Benchmarking & Business Case

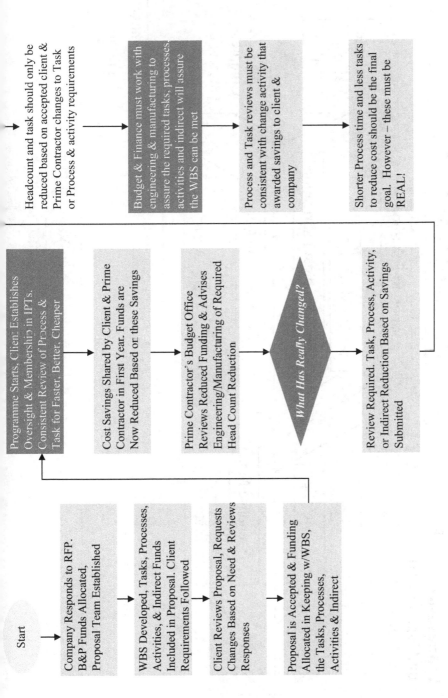

Chart 36   Assessing the Tests, Processes and Activities Needed to Complete a Contract

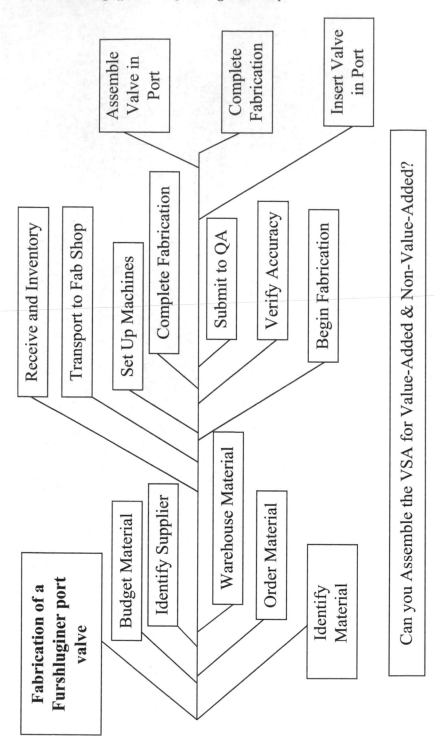

Chart 37 Value Stream Analysis

Sales
- Direct Expenses
_____

= Gross Profits
- Indirect Expenses
_____

= Net Profits
- Tax & Dividends
_____

= Retained Profits

Chart 38   Profit and Loss Statement Elements

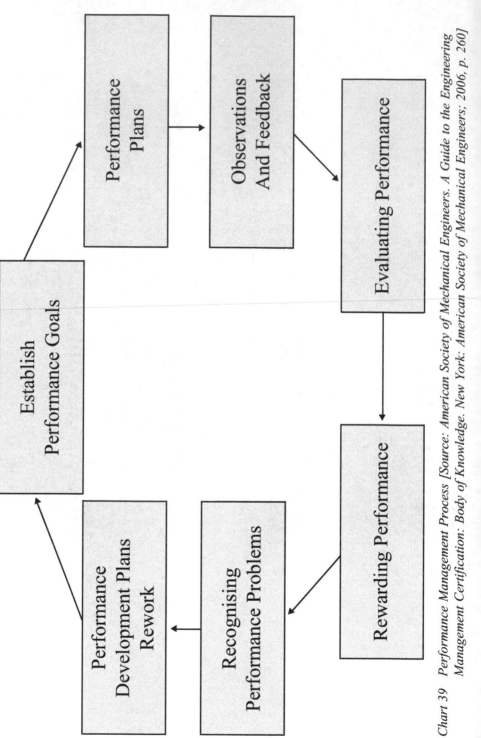

Chart 39   Performance Management Process [Source: American Society of Mechanical Engineers. A Guide to the Engineering Management Certification: Body of Knowledge. New York: American Society of Mechanical Engineers; 2006, p. 260]

- A primary Focus Must be on Teaching & Coaching to:
  - Control Processes & Keep Work Distributed
  - Satisfy the Customer
- Develop Long-range Customer Driven Plans
  - Involving All Employee Inputs
- Devote Time to Studying, Thinking & Learning
  - About Bold New Innovations
  - Concern for the Success of the Company

Chart 40   The Prime Management Function is Leadership

1. Create the Operational Definition for the Company
   - What is the Product, the Vision & the Mission?
2. Develop a Risk Management Strategy
3. Build a Useable Work Breakdown Structure
4. IDs for the Task Relationships
5. Estimate the Work Packages (Time & Cost)
6. Calculate the Initial Schedules
7. Assign and Level the Available Resources

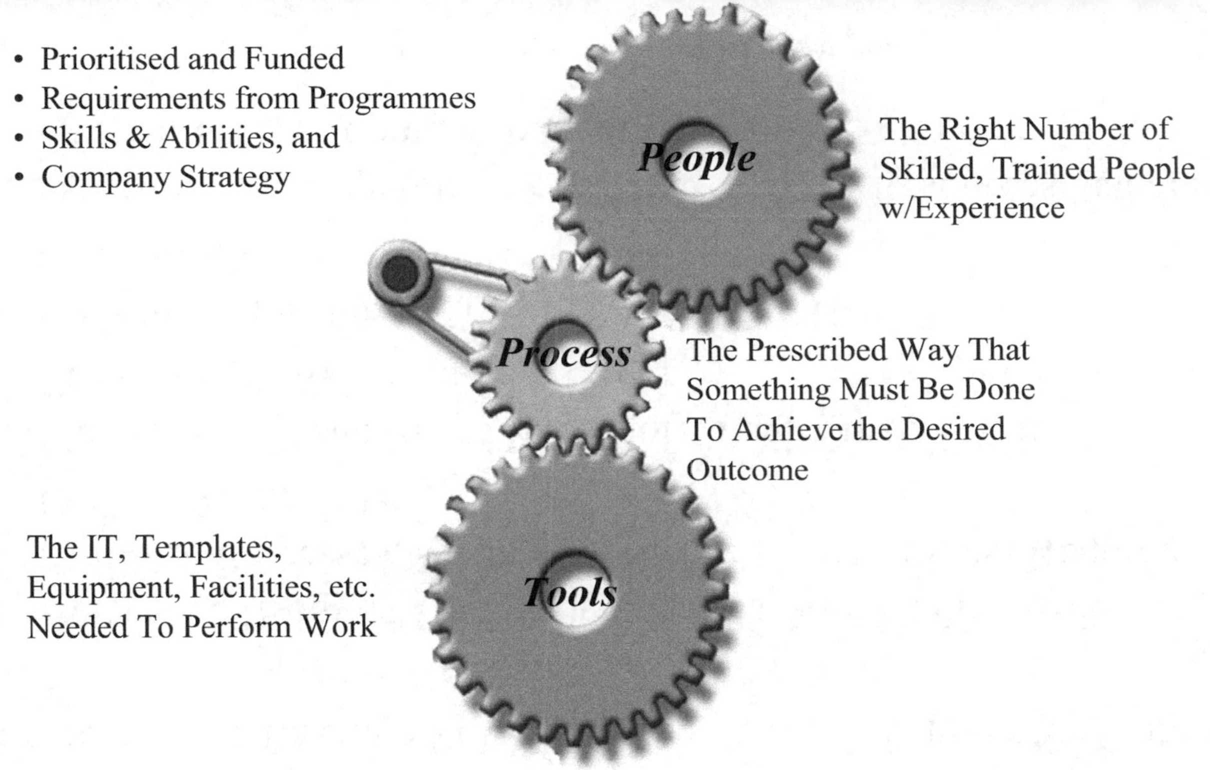

The Right Number of Skilled, Trained People w/Experience

The Prescribed Way That Something Must Be Done To Achieve the Desired Outcome

People

Process

Tools

- Prioritised and Funded
- Requirements from Programmes
- Skills & Abilities, and
- Company Strategy

The IT, Templates, Equipment, Facilities, etc. Needed To Perform Work

A Capability is the Integration of People, Processes and Tools Working Together to Create a Valued Result

Chart 42   *What is Capability?*

1. Identify the Critical Technologies & Processes Used by the Company

2. Identify the Baseline Roles Played by Staff to Support the Critical Technologies & Processes

3. Determine 'Body of Knowledge' (Skills, Knowledge, Attitudes, Abilities, Processes, Methods, Tools & Certifications) Required to Support the Roles

4. Identify the Capability of Each Role, Each Process & Each Technology – Where are the Gaps?

5. Determine the Number of Resources Who Have These Capabilities

6. Determine the Capability Over Time with the Resources that are Available, Developing & Planning to Retire

Chart 43   Eleven Points for Knowledge Development

7. Identify Methods in Place to Maintain Capabilities
   – Mentoring, Cross-training & Internships
   – Programmes, Courses of Study & Re-Training (In & Outside)

8. Establish Policies & Publish to Ensure Action for Critical Needs
   – Performance & gap assessment, quarterly review
   – Management development, skill & knowledge training

9. Establish Change Review Board – Meets Monthly to Enact Change to Policy & Process

10. Eliminate Capability Gaps in Training Workforce or Hire New Employees that Have the Required Skills

11. Maintain Education & Training Programs for New Skills, Changes in Operations & Gap Reduction

Chart 44   Eleven Points for Knowledge Development (cont.)

- Problems begin & continue because no-one teaches the employee how to control or master the processes

- If the employee looks to the manager to solve their problems they will not understand:

  – Variability

  – Supplier – customer links

  – Process improvement

- Teach, coach & guide on a real-time basis

  – How are they doing NOW, instead of once a year

Chart 45   Teaching & Coaching

A Four Panel Status Chart can be used to identify the development/execution of the capability plan

A <u>Accomplishments</u>

- List significant completions performed (eg. milestones, key reviews, critical hardware receipt, etc.)
- List initiative gates accomplished
- Critical cost points reached

B <u>Critical Milestones</u>

<u>Item</u>	<u>Date</u>	<u>Trend</u>

- Status of Critical Stage gates
- Status of future Stages & Completions
- List important Initiatives
- Trend indicators are:
  - O = On track
  - S = Slip from scheduled date
  - A = Ahead of Schedule

C <u>Issues Watch List</u>

<u>Item</u>	<u>Action Plan</u>	<u>Closure Date</u>

- Any Critical Item w/trend of **S**
- List is used to flag non-milestone items
- Recovery plan required to return to **O**

D <u>Milestone Health Indicators</u>

	<u>Current</u>	+3Mos
Milestone #1	Red	Yellow
Milestone #2	Yellow	Green
Milestone #3	Green	Green

Chart 46   Four Panel Status Chart

Key: Blue – Task Complete, Green – Task on Schedule, Yellow – Behind Schedule but Doable, Red – Task in Jeopardy

Engineering - XYZ Corp

Objectives		*Activity*
1. Objective 1	Yellow	1. Activity 1
2. Objective 2	Blue	2. Activity 2
3. Objective 3	Green	3. Activity 3
4. Objective 4	Yellow	4. Activity 4
5. Objective 5	Green	5. Activity 5
6. Objective 6	Yellow	6. Activity 6

Schedule

Tasks	Jan	Feb	Mar	Apr	May	Jun	Jul	Aug	Sep	Oct	Nov	Dec
1. #1	▲		△									
2. #2												
3. #3												
4. #4												
5. #5												
6. #6												

Chart 47   Three Panel Status Chart

Further reading

P-6 ready by crew (1934)

1. Altman, E.I. *Corporate Financial Distress*. New York: John Wiley and Sons; 1983
2. American Society of Mechanical Engineers. *A Guide to the Engineering Management Certification: Body of Knowledge*. New York: American Society of Mechanical Engineers; 2006
3. Associated Press, Rexrode, C., Condon, B. 'CEO Compensation Reaches New High'. *The Atlanta Journal Constitution*. 26 May 2012, pp. A12–A13
4. Bauer, E.E. *Boeing in Peace and War*. Washington, DC: TABA Publishing; 1991
5. Bennet, A., Bennet, D. *Organizational Survival in the New World*. Amsterdam, NE: Elsevier; 2004
6. Bennis, W., *On Becoming a Leader*. New York: Addison Wesley; 1989
7. Berkowitz, B.D. 'The Wall Street Decade'. *Air and Space Magazine*. June/July 1998, pp. 43–7

8. Boeing. *Total Quality Improvement: A Resource Guide to Management Involvement.* Boeing; 1986

9. Bouck, J. 'Unlock the Potential of your Team'. *Automotive Industries.* March 2003, pp. 48–49

10. Browning, T.R., Heath, R.D. 'Lean Implementation Pitfalls in Low Volume Production of Complex Systems: Lessons from the F-22 Program'. Presented at Texas Christian University, M. J. Neeley School of Business, Fort Worth, TX, 2006

11. Cario, P., Dotlich, D., Rhinesmith, S. 'The Unnatural Leader'. *Training and Development.* March 2005, pp. 26–31

12. Carter, R.A., Antón, A.I., Dagnino, A., Williams, L. 'Evolving Beyond Requirements Creep: A Risk-Based Evolutionary Prototyping Model'. *Proceedings of the 5th IEEE International Symposium on Requirements Engineering.* Raleigh, NC: IEEE; 2001, pp. 94–101

13. Columbia Accident Investigation Board. *Columbia, Report Volume 1.* Washington, DC: National Aeronautics and Space Administration; 2003

14. Colvin, G. 'The Great CEO pay heist'. *Fortune Magazine.* 25 June 2001, pp. 64–70

15. Conlon, R., Insler, D., Kochanski, J. *Rewards of Work Study.* Sibson Consulting [online]. 2009. Available from http://www.sibson.com/publications -and-resources/surveys-studies/?id=252 [Accessed 6 May 2013]

16. Conner, D.R. *Managing at the Speed of Change.* New York: Villard Books; 1994

17. Creech, B. *The Five Pillars of TQM.* London:Truman Tally Books; 1994

18. Csikszentmihalyi, M. *Flow: The Psychology of Optimal Experience.* New York: Harper Perennial; 1990

19. Csikszentmihalyi, M. *Finding Flow: The Psychology of Engagement with Everyday Life.* New York: Basic Books; 1998

20. Drucker, P.F. *Management Challenges for the 21st Century.* New York: HarperCollins; 1999

21. Frappaolo, C. 'Consultants View: Building a knowledge management program'. *Beyond Computing,* 14 September 2000

22. Frappaolo, C., Wilson, L.T. 'After the gold rush: Harvesting corporate knowledge resources'. *Beyond Computing,* 16 January 2001

23. Freeman, D.G., Hinkey, M.E., Martak, J.W. 'Integrated Engineering Process Covering All Engineering Disciplines'. Presented at SEI Conference; Pittsburgh, PA, 2002

24. Giddens, D.P., Borchelt R.E., Carter V.R., Hammack W.S., Johnson J.H., Kramer V., Natale P.J., Scheufele D.A., Sullivan J.F., Pearson G., Keitz M., Arenberg C., Ivancin M. *Changing the Conversation.* Washington, DC: National Academy of Engineering; 2008

25. Guderian, B. *Leadership Development and the Role of Continuing Education.* Presented to graduating class of the ELITE Program; Tulsa, OK, 2010

26. Gunn, T.G. *Manufacturing for Competitive Advantage.* Cambridge, MA: Harper and Row; 1987

27. Hammer, M., Champy, J. *Reengineering the Corporation: A Manifesto for Business Revolution.* New York: Harper Collins Publishers; 1993

28. Haughey, D. *Project Planning: A Step by Step Guide* [online]. 2012. Available from http://www.projectsmart.co.uk/project-planning-step-by-step.html [Accessed 26 April 2013]

29. Heymans, B. 'Leading the Lean Enterprise'. *Industrial Management.* September/October 2002, pp. 28–33

30. Hayes, R.E. and Wheelwright, S.C. *Restoring Our Competitive Edge: Competing Through Manufacturing.* New York: John Wiley and Son; 1984

31. Howe, N., Strauss, W. *Millennials Rising, The Next Great Generation.* New York: Vintage; 2000

32. Howe, N., Strauss, W., Markiewicz, P. *Millennials and the Pop Culture.* Great Falls, VA: Life Course Associates; 2006

33. Howkins, J. *The Creative Economy: How People Make Money From Ideas.* London: Penguin Books; 2001

34. Humphrey, W.S. *Characterizing the Software Process: A Maturity Framework.* Software Engineering Institute, June 1987, pp. 1–20

35. Humphrey, W.S. 'Characterizing the Software Process: A Maturity Framework'. *IEEE Software.* 1988;5(2): 73–9

36. Institute of Configuration Management. *CMMII Versus Other CM Certification Programs* [online]. 1988–2004. Available from http://www.icmhq.com/cmii-whitepapers.html

37. Kennedy, M.M. 'Career Strategies'. Presented at ASEE College Industry Education Conference, ASEE, San Antonio, TX, 2007

38. Keyte, B., Locher, D. *The Complete Lean Enterprise, Value Stream Mapping for Administrative and Office Practices,* New York: Productivity Press; 2004

39. Kubler-Ross, E. *On Death and Dying.* New York: Collier Books; 1997

40. Kyd, C.W. 'How are we Doing?', *Inc. Magazine.* February 1987, pp. 121–3

41. Landy, F.J. and Conte, J.M. *Work in the 21stCentury.* 2nd edn. Indianapolis, IN: Blackwell Publishing; 2007

42. Lareau, W. *American Samurai: A Warrior for the Coming Dark Ages of American Business.* New York: Warner Books; 1992

43. Lewin, K. *Group Decision and Social Change.* New York: Holt, Rinehart & Winston; 1958

44. Loomis, C.J. 'This stuff is wrong'. *Fortune Magazine.* 25 June 2001, pp. 73–84

45. McManus, H.L., Millard, R.L. 'Value Stream Analysis and Mapping for Product Development'. *Proceedings of the International Council of the Aeronautical Sciences, 23rd ICAS Congress*; Toronto, Canada, 8–13 September 2002

46. Malan, R., Bredemeyer, D. *Functional Requirements and Use Cases.* 2001 [online]. Available from http://www.bredemeyer.com/pdf_files/functreq.pdf [Accessed 26 April 2013]

47. Maxwell, J.C. *The 21 Irrefutable Laws of Leadership.* 10th edn. Nashville, TN: Thomas Nelson Publishing; 2007

48. Morrison, R., Ericsson, C. *Developing Effective Engineering Leadership.* London: Institution of Electrical Engineers; 2003

49. Ogbonnia, S.K.C. *Political Parties and Effective Leadership: A Contingency Approach*. PhD dissertation, Walden University; 2007
50. Oncken, W. *Managing Management Time*. Englewood Cliffs, NJ: Prentice Hall; 1984
51. QSS Consultancy. *Requirements Management Executive Forum*. Detroit, MI: QSS International; 1998
52. Quality Process Magazine. 'The Golden Rules and Commandments of Business Process Re-Engineering'. *Quality Process Magazine*. December 1994
53. Robbins, S.P. *Organizational Behavior*. 9th edn. Englewood Cliffs, NJ: Prentice Hall; 2001
54. Ross, J. 'Space Needs New National Consensus'. *Aviation Week and Space Technology*. 28 January 2013, p. 50
55. SAE Notes. 'SAE Total Quality Management Process Map'. December 1994
56. Slater, R. *Jack Welch and the GE Way*. New York: McGraw-Hill; 1999
57. Software Productivity Consortium. *Capability Maturity Model, Version 1.1*. The Software Engineering Institute, Carnegie Mellon University; 1991
58. Software Productivity Consortium. *A Systematic Approach to Process Engineering*. Herndon, VA; 1999
59. Tidwell, J.P. 'Industry Speaks with One Voice', Presented at ASEE Annual Conference; Honolulu, HI, 2006
60. Toffler, A. *Future Shock*. New York: Random House; 1999
61. Tufte, E. *PowerPoint Does Rocket Science* [online]. 2003. Available from http://www.edwardtufte.com/bboard/q-and-a-fetch-msg?msg_id=0001yB [Accessed 6 March 2013]
62. Useem, J. 'Boeing vs. Boeing'. *Fortune Magazine*, 2 October 2000
63. Verzuh, E. *The Fast Forward MBA*. 2nd edn. New York: John Wiley and Sons; 2005
64. Vicere, A.A. 'Breaking the Mold: Strategies for Leadership'. *Personnel Journal*. May 1987, pp. 67–78
65. Vicere, A.A., Bitner, S.W., Freeman, V.T. *Management Skills Assessment*, Penn State Executive Programs, 1985
66. Waitley, D.E., Tucker, R.B. 'How to think like an Innovator'. *The Futurist*. May–June 1987, pp. 9–15
67. Wysocki, B. 'Grooming Leaders'. *National Business Employment Weekly*. 30 August 1981, pp. 19–20
68. Zubrow, D. 'Putting "M" in the Model: Measurement in CMMI'. The Software Engineering Institute Conference. Carnegie Mellon University, 2007

Index

Page numbers followed by c indicate a chart.